The Art and Science
of Geography

The Art and Science of Geography

U.S. and Soviet Perspectives

EDITED BY

Vladimir V. Annenkov
and George J. Demko

Routledge
Taylor & Francis Group

LONDON AND NEW YORK

First published 1992 by Westview Press

Published 2019 by Routledge
52 Vanderbilt Avenue, New York, NY 10017
2 Park Square, Milton Park, Abingdon, Oxon OX14 4RN

Routledge is an imprint of the Taylor & Francis Group, an informa business

Library of Congress Cataloging-in-Publication Data
The art and science of geography: U.S. and Soviet perspectives /
   edited by Vladimir V. Annenkov, George J. Demko.
      p.   cm.
   ISBN 0-8133-8516-4
   1. Geography—United States.   2. Geography—Soviet Union.
I. Annenkov, Vladimir V.   II. Demko, George J., 1933–   .
G99.A78   1992
910—dc20   91-38970
                                                                        CIP

ISBN 13: 978-0-367-29024-5 (hbk)
ISBN 13: 978-0-367-30570-3 (pbk)

Contents

Foreword

International comparisons between the concepts and practice of geographers are always difficult. The language barriers and the social and scientific contexts differ so much that it is awkward to assess similarities. Are such similarities real or only apparently real? Is the logic behind the different approaches based on the same principles? In order to avoid mistakes and to generate satisfactory interpretations it is necessary to have some perspective on the sociology of knowledge in the countries that are being compared.

The major goal of this book is to provide explanatory statements about the discipline and, at the same time, to convey a feeling of the differences between intellectual atmospheres in the two countries. The similarities are more numerous than might be expected from nations with such different histories and so few scientific interactions during the past fifty years. Soviet and American geographers have coincident views concerning the development of geography: a significant change occurred during the last generation with the increased use of quantitative techniques and satellite imagery, a growing interest in theory, and more sophisticated modes of explanation. And, there is agreement on the significance of the new methodologies and their impact on the development of the discipline. Some comments, American as well as Soviet, are more cautious: geography will need, in the future, increasingly efficient tools to acquire and to process data; geographers will rely increasingly on their numeracy, but at the same time, progress will depend on an improved capability to decipher processes and to build theories. The significance of the "postmodernist" approach is presented by V. L. Kagansky. On the American side, A. Pred is very vocal when stigmatizing the dangers of over-technical forms of geographical development.

The United States and USSR have operated in very different intellectual contexts during their history and their institutions have few, if any, similarities. What is striking in such a situation is how analogous the positions of American and Soviet geography are within both their national scientific contexts. Geography is experiencing problems regarding its scientific status and scientific visibility both in the United States and in the Soviet Union: the quality of the geographers and of the geography which they practice is, however, not responsible for this situation. The problem is rooted in the nature of geography as a discipline and of its relationship with other disciplines, both in the natural and the social sciences, especially the fact that geography is both a natural and social science.

The final similarity relates to geography in the schools. Differences seem extraordinary: In America, geography is practically absent from the secondary school curricula; in the Soviet Union, geography is taught at every level. But in both countries, school geography is undergoing similar difficulties. The geography taught in both nations is not adequate for today's social needs. It still relies too heavily on the traditional views in the discipline, hence there is a great necessity to promote, in both countries, new curricula and to improve the skills of people in charge of the teaching of geography.

The differences between Soviet and American geography, however, are as significant as the similarities. The concepts of geography differ widely in the two countries. If we accept the idea that geography is made up of two complementary parts, the study of vertical (or ecological) relations between man and environment, and the study of horizontal (or social) relations between humans and between groups, Soviet geographers put more emphasis on the first aspect, and Americans have a more balanced, or even a slightly more social science approach to the field. At the same time, when reading the answers to the question "How is the discipline of geography defined?" one becomes aware of the importance of institutional backgrounds in the way geography is practiced. Soviet geographers rely on a universally accepted paradigm of the core of the discipline: they all speak of the "landscape mantle" of the earth. Their answers convey the feeling that they have been trained in a coherent academic discipline, with clear definitions of its aims and scope. American answers give a different view: as geographers, they have not been trained according to uniform principles; geography is for them a looser construct, and their definitions are more derived from their personal experience in the field and in various schools of research, rather than from a uniformly agreed-upon core.

Differences between the two sets of responses do not proceed only from the different academic environments in the two countries. They also reflect the emphasis given to ecology. Soviet geographers are more interested in ecological problems, although some Americans have similar attitudes: the view of John R. Mather, for instance, is sometimes parallel to the prevailing Soviet ones. In the USSR, some geographers are more aware of the historical and social dimensions of geographical phenomena and thus closer to the prevailing American views: such is the case of Vladimir Annenkov or of V. L. Kagansky, for instance.

Soviet responses provide us with a view of the major achievements of Soviet research in the field of environmental preservation. The statements, especially of V. V. Vorobyov, B. B. Prokhorov and A. M. Trofimov, and R. G. Khuzeyev (as well as others) describe the impact of Soviet geographical research on the preservation of environments and on the management of natural resources. I am not sure that Western geographers are fully aware of the quality of this kind of research and of its implications. Soviet geographers played a decisive role in making Soviet citizens conscious of the problem of their environment: it is an outstanding success!

This interest in natural preservation is also central in K. Ya. Kondratiev's studies on the global consequences of nuclear warfare. It is certainly one of the fields where Soviet geography achieved impressive results.

There is much more in *The Art and Science of Geography: US and Soviet Perspectives* than the expected academic statements on the excellence of the discipline in both countries: The book provides interesting insights into the social and intellectual settings of the two schools, and allows one to see how much they converge and how far they diverge. It is refreshing to see scientists working the difficult path toward a better understanding of their positions and practices. European geographers cannot remain indifferent when confronted with such an evolution. I think I express their views when saying that they must also be in some way instrumental in the East-West dialogue that is now starting in our discipline.

Paul Claval
Université de Paris, Sorbonne

Preface

Geography, one of the oldest sciences on earth, is also one of the least understood by the general public. Many people consider geography to be the discipline that maps locations and describes places and regions. For them a tourist guidebook is a work of significant geographical merit. On the other hand, Aborigines of the Australian desert or Amazon Basin, who are uniquely learned regarding interrelations between nature and man, are considered ignorant of geography. Given such contradictions, we hope that this volume will shed some light on, and bring about greater understanding of the discipline of geography as practiced by individuals in two of the largest geographical communities in the world.

Many professional geographers consider contemporary geography to be a newly developing science with significant potential to influence individual and societal behavior. The art and science of geography accelerated in the 1950s at such a rate that a single generation of geographers witnessed many revolutions in the discipline --- quantitative, predictive, philosophical, ideological, and humanistic. The balance between traditional geography and contemporary, innovative geography has changed and continues to change. For many in the profession it is difficult to keep current regarding advances in the field --- mountains of new data, facts, methods, and concepts are continually generated by various subfields and schools of geography. A proper view of the art and science of geography and new types of cooperative efforts based on international cooperation is required.

The idea for this volume was generated as a result of the Soviet-American Subcommission on Geography deliberations on joint research and publication. The fundamental question addressed by this volume is how contemporary geography, as an art, a science, and an educational discipline is defined and viewed by practitioners in the United States and the USSR. Fourteen questions were formulated and distributed to a select group of geographers in both countries with the help of the Geographical Society of the USSR and the Association of American

Geographers. In the United States, thirteen scholars from a list of forty prominent geographers responded to the questions. In the Soviet Union, answers were received from all forty geographers that were contacted. The List of Contributors and their affiliations follows the text.

The reader will readily note the differences and variations in views between national groups and even within each community. This diversity, more in concept than in approach, reflects weaknesses as well as strengths of modern geography. The nongeographer, observing the diversity of concepts evident in the responses, may conclude that geography may be more an art than an exact science. However, one should understand that our diversity results largely from the transitional stage that our discipline now occupies. Currently, there is competition between different groups and schools of geographers that greatly stimulates the development of research and its application. Much of this intellectual concern and variety is evident in the spectrum of views presented in the following pages.

The editors have worked together on a number of occasions to prepare this volume within the framework of the Subcommission on Geography of the Commission of the Academy of Sciences of the USSR and the American Council of Learned Societies. We are very grateful to both organizations for their support of this and related projects. We owe a debt of gratitude to many colleagues and friends who were helpful in many ways, but especially to Lena Shevchenko of the Institute of Geography in Moscow and to Kerstin Demko for their assistance with translation from Russian to English and vice versa, as well as editing.

We should also note that at the time of publication of this volume, the Union of Soviet Socialist Republics ceases to exist and has been replaced by a number of new, independent states. Such an event is enormously important politically and economically and has significant impacts across the globe. Geography and the way it is practiced in Russia and the new states, however, will not be greatly affected in the short term. Most geographers in Estonia and Ukraine will largely reflect the training provided during the Soviet period. Obviously the pace of change in terms of theory and methods will accelerate with more frequent and intense contact with the rest of the world's geographic communities and the problems selected for study will reflect new state concerns. The Soviet traditions from Baransky, Saushkin, and Anuchin as well as earlier Russian traditions will continue to characterize geographic research and teaching for decades to come in Russia and the new independent republics.

The editors sincerely hope the view offered of the art and science of geography alters the popular and often incorrect image of the discipline and its divisions. Further, we trust that our readers will sense the dynamism, variety, and significance of geography in the USSR and the United States. Finally, we hope that this material will stimulate international discussion on the important trends, needs, and problems of geography as a developing science on the threshold of the twenty first century.

V. V. Annenkov
G. J. Demko

CHAPTER ONE

The Essence of Geography as a Discipline

How is the discipline of geography defined?

A brief preparatory comment regarding this question is necessary. In Soviet geographical science a long tradition separated its subfields by the objects or phenomena studied. The environment and society are generally perceived to be the two main sets of objects for research. Thus, the division between physical geography, the study of natural phenomena, and economic geography, the study of economic and demographic aspects of society, persisted into the Soviet period. The integration of these two branches has become the predominant trend in Soviet geographical science over the last two decades.

In the United States, the integrated nature of physical and human geography date to the early part of the twentieth century and the majority of geographers focus more attention on the various approaches and methods used to study reality. The "field of geography" is defined by Americans usually in these terms and not by the various social or physical phenomena that are studied. Thus, the focus of the answers to the first question by the Soviet and American geographers appears somewhat disparate, reflecting the differences in the two scientific schools.

Pred (USA): As a noted English literary scholar observed a few years ago, it would be a rash person who would define something as basic as geography or philosophy, or distinguish neatly between sociology and anthropology, or advance a view of a field as an unchanged set of practices. Any single-sentence definition of geography can only contribute to division (among the included and excluded practitioners), to the promotion of orthodoxy, to the preservation of a scholarly status quo, and to the repression of intellectual innovation. If it is unrealistic and counter-productive to briefly define our field, one can point to four overarching and interlocking themes that have characterized most, but not all, of the research conducted within it for a long period of time. As the questions, emphases, approaches, and methodologies employed within geography have changed with time, the following four themes have persisted:

- The role of humans in physically transforming the earth's surface.
- The processes by which society organizes space, and the impact of

3

spatial organization, or spatial structure, on the workings of society and the lives of people.
- The activities or humanly created landscape features that characterize areas (regions, cities, etc.) as the product of social, economic and cultural processes operating within the context of given natural systems.
- The interaction of natural systems (climatic, geomorphic, biotic, and edaphic) at the earth's surface.

Willmott (USA): No universally accepted definition of geography exists, although several threads are common among the works of geographers. These include the following:

- Description of the geometric manifestations (on the surface of the earth, that is) of both natural and human processes and phenomena.
- Primary concern for the role of location and distance (where distance may be Euclidian, functional, etc.). It has been said that "distance is to geography what time is to history." Location is implicit in this statement and space is proportional to some power of distance.
- Emphasis on those natural and human processes that are significantly influenced by distance, location, and space. Some of these are geopolitics, geoeconomics, migration, weather, climate, and so on. It is reasonable to argue that geography focuses or should focus on those phenomena and processes that are fundamentally affected by location, distance, or space.

Cohen (USA): There is no single and overriding definition for geography. The various approaches provide their own definitions, for example, spatial relations and interactions; area differentiation; man-environment relations; and earth-surface processes. For me personally, a good definition is that geography is the study of the attributes of the system of place — its content, boundaries, and intrasystemic connections that shape its structure; and its extrasystemic connections that both differentiate the system and feed back into it.

Taaffe (USA): I have never found a single definition to be very satisfactory. I have been using a "three traditions" definition since the mid-sixties and have found no reason for dissatisfaction. These are: (1) the spatial organization view stressing maps and spatial analysis; (2) the area study view, stressing synthesis, integration, and a concern with place; (3) the man-land or ecological view, stressing relations between man and the natural environment.

Pattison, in 1963, used these three together with a fourth view, that of earth science, which is certainly valid historically and currently in many countries other than the United States. I find it fails to distinguish between physical geographers, on the one hand, and meteorologists and geomorphologists, on the other, who are certainly earth scientists. For the most part, work in physical geography is distinguished by its emphasis on interrelationships among physical phenomena (area study or integrative view) or ecology (man-land view) and to some extent on spatial analysis, although sophisticated map use is widespread in the other disciplines.

Palm (USA): The field of geography is defined as the study of the surface of the earth, including its climate, landforms, and biota; this includes the processes through which the surface of the earth is transformed, both its physical character and also the ways in which "places" (settings for human activity) are created and transformed as well as human activity on the earth's surface, and the mutual impacts of this human activity, and the character of the environment. There are two principle questions that devolve from this definition:

1. What is the relationship between people-societies and their environment?
2. What is the spatial organization of human activity?

These two questions form the basis of all of human and physical geography.

Brunn (USA): Geography is a discipline that focuses on location, place, and environment. It studies the variations of human and physical phenomena that exist at the earth's surface, how they came to be (that is, the structures and processes), what they look like (the patterns), and what they mean to individuals and peoples (that is, interpretations and experiences). The phenomena geographers study include those from the human or social sciences and the earth or natural sciences. Geographers are interested in where these phenomena are, how they came to be where they are (the process and time dimensions), and where they might be (a forecasting and predictive inquiry). The scales of inquiry range from the local to the global.

Geography is viewed as a bridging discipline. It searches for and examines relationships among those disciplines and phenomena studied by social and earth sciences. The man-land tradition in our discipline, with its focus on human-environment relations, seeks to understand the interrelationships between the human and earth sciences. Examples include human settlement patterns and agricultural economies, climate change related to energy consumption, and human behavior and environmental perception   (the natural hazard school).

Geography is also a bridging discipline between the arts and sciences.   The humanistic geographers address questions of place and landscape, its meanings, symbols, values, and interpretations.   The spatial geographers look at questions related to the science and geometry of location and space.   They are concerned with the search for meaningful spatial theory, regularities, and prediction in explaining processes and patterns of industrial location, travel behavior, synoptic climatology, regional economic change, and other phenomena.   The humanistic and spatial approaches focus on earth environments and contribute to our knowledge of humans and environments through understanding and experiences (the humanistic geographers) and regularity and predictability (the spatial school).   The common ground is an attempt to increase our knowledge and understanding of what goes on where and why on earth.   Additional common interests are in mapping and the language of maps, the search for spatial order in the environment, and the identification of regions for theoretical or applied studies.

de Souza (USA): Geography is concerned with place. Its subject matter is the world around us and the natural and human phenomena that make up the world's environments and places. Its practitioners describe the changing pattern of places in words and maps, explain how these patterns come to be, and attempt to unravel their meaning. Geography's continuing quest is to understand the physical and cultural features of places and their natural settings on the face of the earth.

Mather (USA): Geography is often defined in terms of a concern for spatial distributions just as history is often defined as concerned with the sequence of events over time.   More specifically, geography is concerned with the interactions of humans with their physical environment on all local, regional, and global scales.   But geography also must consider the earth as a totality and must study the earth as a single system with many complex interactions and feedbacks among the atmospheric, biospheric, and oceanic subsystems that make up the physical system and among humans, their cultures, their economies, their societies, and their own interactions with one another.

Preobrazhensky (USSR): The common and finite object of research in geographical science is the landscape mantle.   This complex, heterogeneous mantle and its component parts (lithosphere, atmosphere, hydrosphere, pedosphere, biota) form an open dynamic system which has spatio-temporal arrangements and organization. Changes taking place in its different parts impact very widely through the system due to the interconnections among its components and over its space. This mantle has gone from an abiotic stage of development to a biospheric stage and has entered a stage of transition to biospheric-noospheric development. Its condition is increasingly dependent on the activity of man and society.

Geography is the only system of sciences that has chosen for its subject matter such a heterogeneous set of systems. Therefore, the role of geography in the understanding of the world is unique.

The geographical character of our research stems from emphasis on the spatial-temporal organization of the objects under investigation and the understanding of place and location in the complex systems of the mantle.

Velichko (USSR): The task of geography is to study the structure of the landscape mantle as an open system with an external energy source, the dynamic relations that determine its organization by the energy-mass exchange between the components of the mantle, relief, vegetation cover, hydrosphere, and so forth, and the evolution of the mantle. Geographers also seek laws relating to rational use of the mantle by man in the process of economic-cultural activity and the spatial distributions that result from this use.

Man is not only a witness to but is the main cause of one of the major aspects of the evolution of the landscape mantle, its transition from a purely natural system into a natural-anthropogenic one, an irreversible process. At present, two groups of factors, natural and anthropogenic, obeying fundamentally different laws, exert a decisive influence on the state and development of the landscape mantle. The range and rates of purely natural change and fluctuation in the state of the landscape mantle are relatively well known, but the anthropogenic factor in the formation of the landscape mantle has been very poorly studied so far. Thus it is natural to advance, as the most important task of modern geography, the development of a new leading scientific direction studying the laws and mechanisms of the anthropogenic impact on the components of the landscape mantle and through them on the landscape mantle as a whole.

Kotlyakov (USSR): The key problem of geography is to expand our knowledge of the general laws of the earth's landscape mantle. This aspect of our science deals with the laws of the structure, functioning, dynamics and evolution of the landscape mantle at different regional scales: global, continental, individual country, and the local level.

The spatio-temporal laws of the interaction of nature and society are still unclear and no adequate measure of the negative and positive results of this interaction has been made. Neither the prediction of changes in natural-anthropogenic geosystems, nor principles for controlling them have been developed. The study of rapid changes of nature caused by the realization of major economic projects is still inadequate by contemporary standards, and the problem of conserving resources and environmental reproductive capacity of geosystems has not been solved.

A new paradigm for earth science is based on the concept of integrating the two main branches of geographical science, physical-geography (studying laws of nature) and socioeconomic geography (examining the laws of the economy and society). This approach predetermines a decisive shift of geography toward the study of the interaction and development of natural, natural-technical, and socioeconomic geosystems.

We are increasingly aware of the fact that the biotic and abiotic components of the biosphere are closely interconnected, that the impact of man on the earth is becoming comparable with natural-historical processes, that the possibilities for man to live on earth are limited, and that modern advances in science and technology make it possible to study the earth as an interrelated system.

We must learn how to detect and react in a timely manner to significant global changes occurring on the earth at present as well as those likely to occur in the near future. The increase in energy demand together with its anthropogenic impact on nature testify to a crisis developing on our planet. These human development processes decrease the size of natural areas (e.g. forests)leading to the deterioration of the atmosphere, soil loss and erosion, and the depletion of natural resources and traditional energy sources. This impact on the economy and pollution of the natural environment affects the natural ecological links throughout the sections of the landscape mantle, finally disrupting the dynamic equilibrium of natural systems and result in many negative global effects. It is only through broad international cooperation that it will be possible to investigate and resolve the ecological crisis on the earth.

Pashchenko (USSR): Regarding the notions geographical mantle and landscape mantle, I prefer the latter as a more complete and accurate definition of the object of geographical research in its scientific and socioeconomic aspects.

The landscape mantle cannot be studies by physical geography exclusive of its anthropogenic (human) component. Reality makes the geographer think not only of an abstract natural object independent of man's influence but also of its concrete modern form with all anthropogenic changes, technogenic inclusions, and diversified economy. The landscape mantle is our biotic environment, containing natural resources and limited possibilities, and because of its vulnerability to pressure from technocracy, its inspiring beauty requires protection.

Kuznetsov (USSR): The aim of research by constructive geography (closely related to what US geographers refer to as applied geography), which unites physical and human geographies, is to determine the best form of interactions of natural (physico-geographical) and social processes. Such modern forms of production organization as the territorial-productive complex (TPC), that is, applying geographical principles and expertise to solve actual problems, is another example of a general object of geographical research. It is in determining such "areas" that the unity of inter-disciplinary links between branches of geographical science is achieved, that is, that interrelations between territorial combinations of natural resources and the structure of the productive complex are thus determined.

For example, erosion as a natural process, while influencing society, affects the socioeconomic sphere. Thus, damage from erosion in the country exceeds billions of rubles, and a conscious and efficient reaction by society to erosion becomes vital.

Annenkov (USSR): With the geographic mantle of the earth (the landscape mantle) as its finite focus, geographers also study local and regional entities as parts of this finite object. Regional studies are conducted in three aspects: observable landscapes (the geographical environment); "invisible" structures, interactions and competing relations among natural regional systems, population, economy as well as within these systems; and the understanding and transformation of landscapes and spatial relations by society.

Let us consider the object of regional geography in terms of an emerging study of the territorial organization of social reproduction [e.g. production of material things from steel to wheat]. There is a general understanding of development as the interrelated functioning of natural, demographic, social-cultural, and economic territorial systems in a long-term perspective, necessary to the renewal of the material objects. Developmental relations are realized at different territorial scales (ranging from local to global), and inside specific territories, transforming the material environment or adapting to it. The development and resolution of contradictions in territorial systems via spontaneous processes or planned regulation by society is the essence of the phenomena of territorial organization of social reproduction.

Turner (USA): Geography is the discipline that focuses on the "why of where" question --- why do things occur where they do? Sometimes this question is important; other times it is not. Geography pursues this question in two fundamental ways --- in vertical and horizontal space, each grounded in pluralism.

The vertical space approach examines interactions of phenomena in places that give rise to the problem or object in question (integrative synthesis). The horizontal space approach examines the relation of the object of study to others located elsewhere (comparative synthesis). In some cases, the two approaches are combined.

These questions and approaches are not exclusive to geography, just as time is not exclusively the domain of history. In practice, however, geography is the only discipline that focuses on both.

Mironenko (USSR): Geosystems, that is, systems belonging to the landscape mantle, where horizontal and vertical links are observed, are the object of geographical research. These may be part of the natural system and part of the social system, but more often they are geosystems where the direct and indirect interactions of natural and social processes take place.

Kagansky (USSR): While studying the surface of the earth, the geographer is interested in the variety of landscapes, both natural and cultural; their specific features; internal and external interrelations; and the laws of the spatial organization of society and its environment. This object set is studied from the perspective, and by means of, its "morphology," the pattern of visible spatial structures. Important aspects of such a study are the ways the societies spatially organize themselves, and concepts that reflect and explain the transformation of landscapes through human actions.

Geographers are constantly situated inside their object of study, and, in order to understand it, gather personal impressions with all their senses together with all the information available to society.

Tobler (USA): Geography is the study of processes that result in changes in the number and arrangement of people on the surface of the earth.

Morrill (USA): The purpose of geographic scholarship is to understand and explain (1) the interaction of humans and the physical environment; (2) how humans create regional and place identity; and (3) human spatial behavior and the spatial structure of society. In other words, how do the facts of space and the differential content of space condition physical and social processes and behavior and result in the observed organization of territory and the differential character of places?

Muller (USA): I define geography quite broadly — what people do on the surface of the earth and why, or the spatial dimension of human affairs. But, it is also important to recognize several traditions: the constant interaction of the earth-science, culture-environment, locational, and area-studies traditions form the cornerstone for any comprehensive statement about the concerns of our discipline. Balance should be stressed, too, balance between physical and human geography, and between systematic and regional geography.

Demko (USA): It has been written that geography is the spouse of history, that is to say, that geography is concerned with the where and why there of things over space, terrestrial and outer, just as history concerns itself with all things over time. In our current world we have often ignored space or locational attributes, much as we have ignored the past. There is great danger in this attitude inasmuch as, just as those ignorant of the past are condemned to repeat the mistakes of history, so are those ignorant of space, location, and spatial processes condemned to error and folly in decisions relevant to location.

The focus of geography on phenomena over space has yielded at least three main research cores that define our discipline. The first of these is a spatial analysis emphasis that centers on questions of why, where, and the processes that bring about and alter spatial distributions. Among some of the more rigorous examples of this research is central-place theory and methodologically, spatial allocation models.

The second research core of geography is the man-land or man-environment emphasis that stresses the analysis of interactions between "human" societal processes and the environment, in all its complexity. In the environmental area we sometimes immodestly refer to the geographers' approach as integrative or holistic. The man-environment focus also has an important spatial dimension, for such interrelationships give rise to uneven distributions at all scales.

The third research core of geography is regional analysis. This approach, one of the earliest foci of geography, is undergoing a strong resurgence in recent times. Its concerns range from the description of unique places, to the process of grouping similar places or locations on various criteria, to the analysis of regional systems, and processes of regional interdependence.

Clearly, the three cores of geography are not mutually exclusive but are interrelated. One of the disciplines' main problems, however, is to convey to nongeographers that such things as place geography and simple description of places are like multiplication tables in mathematics and the periodic table in chemistry. Such simplistic geography is merely the rudiment of an art and science and all too often the form emphasized by our journalistic and pedagogical colleagues.

What are the basic methods and techniques of contemporary geographers?

Most American geographers emphasize that the entire range of scientific methods must be used as appropriate. Almost all agree that the map and cartographic techniques are quintessentially "geographic methods." The majority of the Americans also emphasize the importance of the new tools of science --- computers, automated cartography, satellite imagery, and remote sensing and related technological developments. Similarly, Soviet geographers identify the wide and varied array of methods that geographers may use and also emphasize the role of maps and cartography.

Muller (USA): Methods and techniques fall into broad categories, all governed by careful and logical scholarly investigation, wherever possible conforming to the scientific model. Among the broad groupings are quantitative methods, cartographic methods, and field methods. Computerized applications greatly enhance these methodologies today, but they still rely upon careful interpretation by the trained analyst. New technologies, such as remote sensing, further expand our capabilities. Good writing is a basic method of communication that has become badly neglected in recent years. There is no substitute for cogent, clear writing, and it is vital to keep these skills at a high level in geography throughout the world.

Turner (USA): While it is tempting to argue that certain methods and techniques pervade geography, more so than any other discipline, I am not at all convinced that this is the case. Method and technique are as wide-ranging as are the subjects addressed and philosophies and approaches used. One exception exists. Because of the central question of the discipline and the two fundamental approaches to synthesis, visual expression of space is typically important for all "real world" studies in geography. Therefore, cartographic interpretation is the only technique or method that might be truly that of the geographer.

Tobler (USA): The methods are the same as in any scientific field. Logic, mathematics, exposition, speculation, observation, mistaken ideas, education, and so forth. And, reliance on a tradition of work transmitted by other scholars through refereed literature.

Taaffe (USA): For the most part, geographers share with all scholars a common body of analytical, archival, and survey techniques. To some extent these vary among subfields. Climatologists and meteorologists share many techniques as do economic geographers and economists or cultural geographers and anthropologists. There is very little that is exclusive to geography although we do put greater emphasis on some approaches than do other disciplines. The map is, of course, central to the spatial view and is almost equally fundamental to geographic work based on the area study or man-land view. It should be noted, however, that several of the physical sciences such as geology and meteorology make extensive and effective use of maps whether on paper or computer display. Mathematical models explicitly designed for the analysis of spatial distributions are used much more widely in human geography than in economics or other social sciences, as well as in those aspects of physical geography that are strongly spatial. Such models also have great potential utility in area study and possibly ecological work, although, by definition, man-land work may be carried on without reliance on anything spatial. Mathematical and statistical techniques involving many variables are of considerable potential utility in the integrative or synthesizing aspects of area study. Field work is particularly useful in area study and some man-land work, and survey techniques are important in geography as they are in most of the social sciences.

Voronov (USSR): Modern geography possesses a wide spectrum of methods. The main one is the comparative-geographical method that allows phenomena to be evaluated quantitatively. Thus, in evaluation of the biomass and production of a vegetation type and its animal population, it is the quantitative character of both these indicators that makes it possible to find out how much organic matter is concentrated in a particular community and what part of it can be taken by man without damaging herds. Annual transmission of solid particles by water and the quantity of snow drifted by wind are also evaluated quantitatively.

Evaluation processes are dynamic in character and timed for certain periods: days, month, year, century. It is only such spatial and temporal comparison that allows us to form an idea about the geographical adequacy of different phenomena and processes and about differences in intensity. Thus, the spatio-temporal approach is the basis of both modern and traditional geography.

The map is a necessary instrument of this approach. Without the map the picture of the world cannot be understood. It determines the scale at which the geographer works. The degree of detail of the phenomena considered is related to the scale. The level of studying a particular phenomenon also depends on the extent of the development of cartographic methods. For example, one of the less developed branches, namely, is animal population mapping. Even now it is still frequently the case that zoogeographical maps depict, in addition to regional boundaries, symbolic images of animals denoting certain species. Recently, different animal population maps have been developed, using patterns, that show combinations of animal populations and their habitats. Such maps effectively depict all animal species of a particular region as well as prevailing species and secondary and tertiary species.

Palm (USA): The basic, common analytical tool of geography is, and always has been, the map. What is new to contemporary geography is the number of ways in which maps are produced, and most especially the ways in which information about particular places is obtained and translated onto maps. Computerized cartography, using geographical information systems to put together a potentially vast array of variables about a place into a readable format, and the gathering of data about places through the use of satellite imagery as well as through more traditional ground methods, have greatly expanded our ability to understand patterns and processes on the earth's surface. All of these techniques are combined with standard statistical analysis, and also with a smaller set of statistical techniques particularly applicable to spatial analysis (e.g., spatial autocorrelation) in order to deduce patterns that are visible and immediately obvious, as well as those that require more sophisticated data transformation.

Innovations in statistical analysis, particularly the applicability of uncertainty theory, the analysis of cataclysmic change, and bifurcation theory have important implications for both physical and human geographic analysis.

Trofimov (USSR): Sociological, ecological, and integrational tendencies in geography reflect a characteristic period in the development of geography gradually and in response to global changes, and especially beginning with the 1960s.

Geographical science seems to have received the greatest stimulus from the "quantitative" and "philosophical" revolutions. The tendency to use quantitative methods did not arise in a vacuum. The fundamentals of the formalization of geographical concepts were laid by outstanding traditional researchers. The striving for clear thinking in geography, for its methodological compactness and integration of knowledge, has naturally and firmly converged with the general scientific method, systems analysis, and so forth. The application of quantitative methods has proved natural in these conditions, and they have joined the arsenal of basic approaches in geography.

The quantitative revolution and extensions (i.e., philosophical revolution) meant, also, a transition to a certain ideology and consequently interest in theoretical geography has increased. It is in this respect that geography acquired the features of a modern science capable of solving complex tasks by means of the most up-to-date methods, including mathematical, many of which have been improved or created within geography. Methods of spatial statistics have developed: spatial autocorrelation, diffusion of innovations, heuristic maps developed and used (e.g., statistical surfaces, trend response surfaces). Conceptual problems were posed and related to the functional treatment of unified spatial structures, implying relationships between geographical structures and hypothetical spatial processes. Such approaches opened new possibilities to model multilevel hierarchic systems.

The mathematical orientation, based on the foundation of a systems approach, has become an essential part of geographical knowledge and has led to a special subfield-mathematical-geographical modeling.

The further development of mathematical-geographical modeling on a substantial basis has resulted in the development of new, geosituational analyses. The latter has great possibilities for practical application.

Voronov (USSR): The role of mathematics in the development of natural sciences has been always great. Statistics have been of great importance for geographers for some time. Moreover, new statistical methods were developed, to a considerable degree, by economic geographers and economists. The advent of computers has widened the possibility of applying mathematics and introduced labor-saving methods of using multidimensional statistics.

Unlike mathematical methods, the systems approach widely used in different sciences is based on the fact that any object studied may be examined at a number of levels, each of which is analyzed by a particular method applicable to each level. Thus, different methods should be used in studying the biosphere as a whole, individual zones, communities, as well as their components.

Silenko (USSR): The systems approach appears to have become the main method in modern geography, whose language can be used to describe all other methods. The systems approach may be present in the subconscious, intuitive form in the process of research.

The systems approach can be applied to both qualitative and quantitative methods. Thus, it is widely applicable but should be cautiously and carefully used in research problems. In mathematics, philosophy and other sciences, the systems approach is at a more developed stage than in geography.

Shcherban (USSR): Modern geography widely uses field observation and laboratory experiments for direct and indirect evaluation of the parameters of the changes of natural and anthropogenic landscapes and their elements. The combined use of air, spaceborne, and ground instruments and installations make it possible to obtain objective characteristics for regions at different scales, in different aspects, with different levels of detail and completeness.

Annenkov (USSR): As geography enters the stage of forecasting and programming future environments and human activities, the spatial approach, traditional to geographers, is developing into a spatio-temporal approach.

Two components may be distinguished in this approach: the study of the current functions of territorial systems ("time-geography" of T. Hägerstrand) and the historical-geographical investigation of changes in landscape and the study of structural links between heterogeneous systems.

The study of development perceived as the changes of spatio-temporal structures, is based on the coherent solution of such tasks as:

1. reconstruction of the state of material objects and functioning processes in the past at some temporal cross-sections;
2. the identification and study of the stages of development of an object (process) in different regions of the modern world;
3. the deciphering of latent structures in the light of their history (structural-genetic approach);
4. the study of the successive changes of structures over time and the laws of transition from an old to a new structure;
5. the projection of the processes of development and future structures.

Each of these approaches uses particular methods of research, many of which are quite labor intensive.

Willmott (USA): The methods and techniques of geography include:

1. Maps, air-photos, and remotely-sensed-image interpretation and analysis
2. Cartography, especially digital methods
3. Spatial analysis, including Geographic Information Systems (GIS)

    a. spatial statistics
    b. numerical modeling of spatial processes
    c. mathematics, especially plane and spherical trigonometry, vector and matrix methods
    d. computer programming, especially for cartographics, spatial analysis, and GIS
4. Experimental design and survey techniques
    a. spatial sampling techniques
    b. field methods
        1. physical geography: surveying, soil analysis, microclimatic instrumentation, and measurement
        2. human geography: questionnaire design and use of census data

Kagansky (USSR): The diversity of methods and approaches is easier to characterize if one divides them into groups according to types of geographical research. To denote these groups, architectural-- technical terms have been used rather arbitrarily.

1. Classical (preference for field work and the importance of accurate data, and the presentation of results as descriptions and maps to identify unique regions).
2. Constructive (research for the solution of applied problems, oriented to prediction and the practical use of results and in close association with society and its institutions to evaluate resolve problems).
3. Modernist (quantitative, sciences, using mathematical methodology and research techniques stressing statistical generalization of unique phenomena, assuming the possibility of modeling geographical reality, and developing useful theories).
4. Postmodernist (extension of the classical tradition enriched with advances from other sciences and a general scientific methodology; use of theory to explain concrete problems; accepting geographical professionalism as an important component of the personality of the researcher and the general culture).

These styles are represented in almost all branches of geography, the latter representing the most flexible, developed as a reaction to the fear of the loss of the classical tradition.

Brunn (USA): The current methods include cartography; field and survey methods; library and archival methods; spatial statistics and quantitative methods; aerial photography; remote sensing; qualitative methods; and artificial intelligence. Most younger geographers will have received training in computer cartography, geographic information systems, and multivariate techniques. Those with interests in cultural geography will likely have training in the field techniques in addition to foreign languages. Geographers in the spatial school are likely to be trained in quantitative methods, computer programming, mathematical statistics, geographic information systems, artificial intelligence, and mathematical modeling. Historical geographers are trained in library and archival methods used by historiographers. There are geographers trained whose research interests are mainly in geographic techniques, be they cartography, remote sensing, aerial photography, geographic information systems, or artificial intelligence. Aside from techniques, many geographers are introducing new methodologies to solve geographic problems and study environments, places, and peoples. The logical positivists (science-oriented geographers) co-exist with humanistic geographers, Marxists, structuralists, and regionalists.

de Souza (USA): The methods geographers use range from functionalism, cultural materialism, and structuralism to eclecticism. In modern American geography eclecticism, which some geographers argue is no method at all, is at the forefront.

Geographers combine their concern for environmental and spatial relations with a number of techniques. Their techniques include the use of quantitative methods, computer graphics, and remote sensing. Sophisticated techniques have helped geographers immensely to interpret the world around them, but they are not a substitute for the traditional technique of geography. The most basic geographical technique is fieldwork, which involves seeing and thinking about what is in the landscape. The excitement of observing physical and cultural features in the landscape attracted many of us to geography.

Grigoriev and Kondratiev (USSR): During the first decade of the space age (1957-1967), research from outer space affected meteorology, hydrology, and oceanography and helped solve specific applied problems. The next period (1967-1977) was marked by the launching of manned spacecraft (Soyuz, Apollo), long-term orbital stations (Salyut, Skylab), and automated satellites (Landsat, Meteor-Nature). The progress in space science was accompanied by the development of earth-based research programs and earth science. During this period space information was used in almost all branches of geography, from geomorphology and economic geography to geobotany and zoogeography.

Space observations of earth revolutionized geography in terms of the volume and efficiency of information on different phenomena, the availability of dynamic data series for entire systems at multiple scales, and the comparative analysis of natural and anthropogenic phenomena.   These data, as well as the visual observations of cosmonauts, are used in geography in three main directions.

First, these data are important for studying natural phenomena, their structure, spatial, and dynamic features. For example, it was possible to determine for the first time the real dimensions of dust drifts that can extend over hundreds and thousands of kilometers. Data on the stream structure of some dust formations were obtained and the first sketch map of the foci of powerful dust drifts for the northern hemisphere was compiled. Specifically, the foci of dust drifts on the northeast and southeast coasts of the Caspian Sea were located and a new large focus of storms was found on the northeast coast of the Aral Sea.

The second direction is the study of natural resources. Space survey data are increasingly used in the qualitative and quantitative evaluation and inventory of resources based on increasingly more detailed and effective mapping. For example, in order to evaluate earth resources on a global scale, Landsat images are used that cover almost all the earth's surface. From them, global maps of land use at 1:1,000,000 scale have been compiled. Thus far, however, mineral resources have been rarely discovered using such data. However, new geomorphological maps are now compiled based on space data and, by means of various geomorphological and landscape indicators, natural resource prospecting is enhanced.

Finally, the third use of space observations is linked with the study of human pressure on the environment and nature conservation. Observations from outer space revealed the pollution of water areas (e.g., the southern part of the Baltic sea), pollution of the atmosphere (in particular, from forest fires in 1971 and 1972 in the European part of the USSR), and the destruction of vegetation cover as a result of forestry practice (e.g., around Paris, Leningrad).

Space surveys revealed such problems but also permitted prediction of their spread, dynamics, and trends. For example, the progressive desertification of a number of regions of the world, and in particular, the Sinai Peninsula, was revealed by observations from space as being caused, to a large degree, by destruction of the vegetation cover, and the resultant deflation of soils. Similar desertification caused by man, although on smaller scale, was discerned in the sand deserts of Central Asia. Undoubtedly, such data will be useful for development of nature-conservation measures for entire regions.

Rodoman (USSR): Geography is a science identified in terms of method rather than object of study. For each phenomenon, geographers consider its territorial structure and its external territorial links. Even very small, point-size objects may be geographical if they are included in the system of territorial interactions. Geography accumulates spatial information and expresses it by means of maps and related cartographic methods including satellite images. Maps depict discrete reality as transformed by the verbal-conceptual apparatus of the geographer. The highest level of generalization of such data is regionalization which, in geography, plays the same role as periodization in history, and classification in biology and mineralogy. Nongeographic sciences also make use of regionalization and maps of the earth's surface, but geography is the main generator of these approaches and methods, retaining it by tradition.

Grishankov (USSR): The most commonly made methodological mistake is basing theoretical constructs mainly on a certain method with little regard to the total arsenal of modern scientific methods available. Lately, geography has suffered from the adoption of a number of fads. They are particularly vivid in the attempts to develop geographical theories based on mathematical methods, and, more recently, systems approaches.

An important recent achievement in the development of geography was the gradual identification of general methodological research approaches. These are mathematical, ecological, historical, geographical, and comparative methods. Similar methods may play different roles at varying methodological levels. For example, the systems approach, as a comprehensive dialectical concept makes it possible at a general methodological level to discover holistic properties of an object and to establish internal and external links via functional laws. At the empirical, scientific level, cybernetics may be used to evaluate functional links quantitatively. Now geography is developing its own comprehensive-methodology.

Pred (USA): While computer cartography and remote sensing have some undeniable, fundamental importance in presenting and analyzing data in geography, they represent a threat to the future of human geography, for they inhibit the generation of scholarship that will capture the attention and respect of scholars in other disciplines.

Method follows from theory. And, whatever disclaimers some may wish to issue, all human geography springs from theory. Through their selection of categories and emphases, even the most vehement opponents of theory-informed human geography cannot avoid building their scholarship upon an implicit theory of how the world works in a given setting. Yet the connections between theory and method are, for the most part, consigned to an unexamined discourse. More precisely, to focus on computer cartography and remote sensing is to perpetuate the empiricist stance that has characterized American human geography both before and after the "quantitative revolution"; to limit the real to the visible and measurable; to reject the joint theoretical problematizing of spatial structure and social structure; to shut off methodological considerations which spring from assigning equal reality status to the power relations and rules of behavior that constitute social structure; to inhibit the development of methodologies based on the interplay between the production of regions, places, and landscapes, the production of history, and the biographies of actual people; and to spurn social theory and the dialectical intertwinings of geographically situated practices, power relations, and knowledge of consciousness.

At the same time, to focus inordinately on computer cartography and remote sensing is to become preoccupied with the analysis and interpretation of maps and images rather than with the flux of human activity on the ground; to verify our down-to-earth subject matter; to be ignorant or disdainful of the essential methodological issues associated with field work and archival study; to perpetuate field and archival work that proceeds by methods that are ad hoc, philosophically naive, and not explicitly informed by theory and thereby to perpetuate scholarship incapable of making an impression on those beyond the bounds of the discipline. Until American geography chooses to give at least as much emphasis to the social, theoretical, and philosophical dimensions of methodology, as it does to computer cartography and remote sensing; until it places the methods and techniques of field work on an equal footing with those of computer cartography and remote sensing; until it creates as many academic posts for social theorists and field methodologists as for computer cartographers and remote sensing experts, then it will continue to follow a path that may very well lead to self-destruction.

Demko (USA): The main methods of modern geography are both traditional and new. Given geography's concern with location, the map and cartographic methods are fundamental tools of our field. Spatial patterns have been and will continue to be displayed on traditional maps but also displayed in various forms via computer-cartographic systems. Geographic Information Systems (GIS) have already affected the geographers' ability to retrieve, analyze, and display spatial data at wondrous speeds with incredible creativity.

In addition to cartographic tools, the discipline has developed and continues to develop and adapt statistical techniques, mathematical models, and related methods of analysis for spatial data. Satellite imagery, remote sensing techniques, and systems-analysis concepts are commonly used by our scholars. Hopefully, we will continue to develop our more general skills --- logic, field observation, and, very important, the ability to communicate in written and oral forms.

What is the relationship between geography and other sciences?

Among both Soviet and American geographers there is strong agreement that our field plays a special role among the sciences — an integrating and synthesizing role. That is, geography deals with the subjects of other sciences as a holistic set of spatial processes as society interacts with the environment and each alters the other. In addition, both groups interviewed stress the special role of geographers as spatial analysts of phenomena such factors as soil, culture, and economic phenomena. A number of Soviet and American geographers also describe the special ties of geography with other sciences, such as economics, physics, and mathematics.

Zhekulin (USSR): The interaction of geography with other sciences has deep historical roots. In the past, geographers-explorers collected information on nature, population, and economies of various areas of the earth and thereby contributed to the formation of such sciences as zoology, botany, meteorology, and ethnography. Later, the reverse process occurred, and biogeography, pedology, geomorphology, and climatology evolved. The recent stage of the development of geography related to the ecological and social emphases in science has led to new interdisciplinary ties to biology, human ecology, sociology, and economics.

Geography has two sets of scientific links: the natural sciences and social sciences. This is reflected in the structure of geography. Several structural levels can be distinguished in the organization of science. The first level is differentiation, which includes both natural sciences (geomorphology, climatology, hydrology, pedology, zoogeography, phytogeography, etc.) and the socioeconomic sciences (industrial, agricultural, transport, cultural, and other geographies). The second level reflects synthesis. Natural sciences include earth science, landscape science (i.e., the geochemistry of landscape, the geophysics of landscape). Socioeconomic sciences are represented by socioeconomic regional geography, population geography, the science of territorial-productive complexes. The third level is one of integration. It includes complex geographical disciplines: geoecology, resource analysis, recreation geography, medical geography, the science of natural-economic regions, and historical geography. The fourth level is the philosophy and methodology of geography. It includes such subjects as theoretical and mathematical geography, regionalization, and so forth. The link between sciences at different levels with nongeographical sciences is evident. Thus, the sciences at the first level usually have two sets of connections: one is geophysical, geological, biological, and economic, whereas the other is geographical. This strengthens ties, contributes to the introduction of new research methods (mathematical, geophysical, and biological) and the formation of new disciplines at the "interfaces" of the sciences.

At present, due to the complexity of scientific knowledge, both geography as a whole and each geographical subdiscipline interact with a great number of very different sciences. The number of such "contact" disciplines is about a hundred.

The most frequent contacts for geography at present and in the future appear to be with the systems of biological and socioeconomic sciences. This results from the need to study a wide range of ecological and resource problems as well as regional development problems.

Humankind should be the criterion by which we assess the significance of scientific research. We cannot, however, weaken ties with abiotic sciences; they enrich geography with methods making it possible to evaluate the existence and development of biological components objectively and, primarily, the interrelated life forms --- plant, animal, and man.

Geography has enriched science with many concepts now used by other disciplines. They include the concept of natural-territorial regions, productive-territorial complexes that were adopted by the biological and economic sciences (these concepts have been largely avoided in socialist countries). The study of the effect of the interactions of society and nature is of general methodological importance. Geography also played a role in developing the geomethod (B. M. Kedrov), a specific technique of cognition based on the structural approach, that is, discerning the ties between phenomena, space, and location. This method is unique.

Prokhorov (USSR): The main method in geography, the comparative analysis of different territorial complexes and the extensive use of cartographic techniques, proved to be extremely productive in studying spatially distributed phenomena. As a consequence, geography has close ties with other sciences. Take, for example, some extremely different scientific subjects and spheres of practical activity such as microbiology, philology, military science, economy, demography, and oncology. All are directly tied to geography. The microbiology of landscape is progressively evolving. Researchers have established that each natural complex has its peculiar "microbial landscape." The study of microbiological features of different landscapes also has great scientific and economic value.

Regarding philology, one can find, for example, an atlas of the dialects of Byelorussia created by specialists in dialects and linguistic geography. The *Atlas of the USSR* also has a map showing what languages are spoken in different regions of our country. The interests of military science and geography intersect in the field of military geography, and similar relations are found in industry, demography, and other areas. Recently, an atlas depicting the spread of oncological diseases in the countries of the CMEA (the now defunct Council of Mutual Economic Assistance) has been published in the USSR. This atlas depicts cartographically regularities discovered earlier during special expeditions studying population and analyzing medical statistics. The maps in the atlas reveal that each region is characterized by a specific incidence of disease, for combinations of different forms of cancer. This is obviously of great practical importance for health organizations. The list of sciences connected with and directly intersecting geography is very long.

Some sciences have evolved over time from the unique stem of geography, namely, geology, geophysics, hydrology, oceanography, and climatology, but have retained their ties with geography. Medical geography emerged at the interface of medicine and geography. The old union of geography and botany resulted in geobotany, and geography and zoology gave birth to zoogeography. Geography's role in the development of scientific realms is great. Regional planning plans and projects for various regions, though created in institutes of urban planning are based primarily on geographical concepts.

At present, the most promising research lines appear at the interface of several sciences. Therefore, it seems very important to note the role of geography in the development of a number of contemporary, complex disciplines. In this connection, human ecology is most important. In the development and establishment of human ecology, the United States (Chicago), French, and Soviet schools of geography have made significant contributions. In our country, the first conference devoted to the problems of human ecology was held at the initiative of geographers in 1974 at the Institute of Geography of the Soviet Academy of Sciences.

Brunn (USA): Most economic, social, and political geographers will consider themselves social scientists in that they utilize the scientific method of inquiry to explain the relationships they uncover or discover and the strengths of those relationships. Likewise, most earth science (physical) geographers, that is, the climatologists, geomorphologists, and biogeographers, would be trained in the scientific method and utilize the problem-solving method of inquiry to explain processes and patterns. The science-oriented geographers (or the logical positivists) will have received training in quantitative methods in geography and likewise appropriate numerical courses in geology, meteorology, psychology, sociology, or economics. Both inductive and deductive methods are utilized, although the larger volume of research utilizes deductive methods of inquiry. To many geographers considering the discipline as a science, it is the geometric properties of space that concern them, such as distance, direction, accessibility, size, shape, volume, and interactions. What "scientific" geography contributes to the social and earth sciences is an understanding of the importance of space and spatial relationships in studying phenomena.

Taaffe (USA): In general, there is a continuum between the various topical subfields of geography and the spatial or ecological margins of related disciplines. Basic economic theory is at the core of economics, for example, and as one moves toward geography there is a sequence from regional economist to regional scientist to locational analyst to theoretically oriented economic geographer to regionally oriented economic geographer. From the core of physics one moves through meteorology, process oriented climatology, regionally oriented climatology, and so forth. Inevitably there is overlap at the margins but this, in turn, facilitates the exchange of ideas between disciplines. Similar problems are dealt with by geography and other disciplines, but this permits us to bring different perspectives to bear on complex problems. The existence of a continuum makes geographers less reluctant to apply theories and methods from allied disciplines to their own research if these theories are spatially applicable or can be related to area study or have ecological implications.

Morrill (USA): Most sciences are focused on phenomena; for example, individuals (psychology), groups (sociology), exchange and production (economics); power (political science); culture (anthropology), and so forth, while geography focuses on a dimension (space). Thus geography is both a physical and a social science. It is justified as a distinct discipline because basic spatial processes (diffusion, interaction, locating) occur across the phenomenon-oriented sciences.

Muller (USA): Geography is at the center of a "sunflower," with each "petal" representing a neighboring natural or social science. Part of each intersects the central core, and here we find an overlap of mutual concerns, though our spatial perspective provides us with a somewhat different view than the methods utilized by scientists in the cognate field in question.

Pred (USA): As a consequence of its overarching and interlocking research traditions, geography occupies an area that is overlapped by the natural and social sciences as well as the humanities. While the broad conceptual relationship between geography and other fields is readily apparent, the relationships occurring in practice are all too one-sided. With a few prominent exceptions, geographers are all too prone to borrow models, methods, and materials from other disciplines, and all too little inclined to contribute to the discourse of those fields. Yet now more than ever before, the time is ripe for direct engagement with others. The classical themes of geography have come of age as a consequence of conditions prevailing in the contemporary world. For example, because of hunger in Africa, water supply difficulties in California, the consequences of Amazon Basin deforestation, and a host of other environmental issues, a growing number of scholars outside of geography are attempting to come to grips with the role of humans in transforming the earth's surface. In addition, human geography has come of age in the sense that its theoretical and empirical discourse has been recognized by leading European theorists and social philosophers as essential to the discourse of agency and structure, individual and society, and the nature of power relations, since the social and historical are always socially and historically constituted.

Preobrazhensky (USSR): In a time of scientific and technological progress, when the study of problems rather than phenomena has highest priority, especially interdisciplinary problems, there is more interest in the sciences enriching each other. The ability to work in common scientific fields, rather than in rigid boundaries by discipline, becomes essential.

For example, the problem of man's interaction with nature is a problem typical of interdisciplinary problems. The improvement of geographical research methods with other physical and chemical methods, techniques of remote sensing and computer techniques is trivial. Looking at the question differently, what can geographical science give other disciplines to help solve problems of the man-environment interaction?

First, it can contribute a number of conceptual models. These include:

1. the concept of the landscape mantle as the area of interpenetration and interdependence of the lithosphere, atmosphere, hydro-hydrosphere, biota, and humanity

2. the concept of geosystems, specific types of landscapes as subsets of the landscape mantle with definable properties
3. the concept of noospherogenesis; that is, the process of the transition of the landscape mantle from the biospheric stage to the noospheric stage
4. the concept of geographical chain reactions in which an impact on one of the components of nature has consequences in another far removed from the original point of impact
5. the concept of complex geographical space society and the individual uniqueness of processes at every point in space.

Chemistry, physics, and biology are very important in solving problems of the interaction of nature and society, but, one should not forget that such interaction takes place not in a test-tube or in the integrated circuits of computers but on earth, on the landscape mantle with all its complex and contradictory links between the living and nonliving, the biological, and the human. The models of geography, in spite of all their imperfections, are the main instrument making it possible to apply and synthesize at the earth level the knowledge of physics, biology, economy, and sociology.

The phrase "Humanity has stepped from the age of physics to the age of biology" has become trite. The age of geography is coming!

Mather (USA): Geography has been described as "the queen of the sciences." In recent years, we have not fully accepted that role, preferring rather to emphasize a role only in the social sciences and abdicating to others our role in the natural or environmental sciences. As we more fully develop the idea that geography is concerned with the earth and humans as an integrated system, and as others recognize that only geography has that integrative viewpoint, we will regain our rightful place as the discipline that joins both the natural and social sciences in the search for a better understanding of the world we live in. The individual natural and physical sciences must all contribute importantly to geography, but it should be geography that integrates and develops the "holistic" view.

Turner (USA): Geography is an integrative field among the earth sciences, social sciences, and humanities. Most disciplines in these groups examine problems that involve things with location, and hence a geographic component exists. The linkage with the other groups is not, however, within the abstract domain of location. Geographers have rarely boded well with the abstract theory of location --- a field marked by theoretical economists, physicists, and so on. The integrativeness of the discipline is, in its real-world examination of the event or theory, a situation that invariably demonstrates the complexities and, hence, the need to synthesize the multitude of variables, any number of which the other disciplines fail to consider in detail.

de Souza (USA): In order to describe, sample, measure, and explain what is present on a part of the earth's surface, geographers must refer to knowledge and insights derived from the sciences and arts. Geography is, therefore, a synthesizing discipline. It is a unique position to demonstrate the relatedness of all knowledge.

Increasing specialization in contemporary American geography is a weakness rather than a strength. Many articles published in our journals, and papers presented at our professional meetings are inadequate dabblings in other disciplines. A discipline is known for its themes and conceptual frameworks and not its borrowed techniques and ideas.

The relationship between geography and other disciplines is, therefore, a great worry to distinguished American geographers. American geography needs a larger pool of talent and a greater concentration of talent in a few university geography departments to create recognition outside the discipline.

Demko (USA): The relationship between geography and the other sciences is somewhat complex. Like history, we are not defined by the things, objects, or phenomena that we study, but rather by our concern for the questions of where and why there? Thus, geography is both a social science and a physical science, a humanity and a science, a theoretical and an applied field. The "holistic" and universal nature of geography has been a major problem in conveying an appropriate and accurate image to nongeographers, other scholars, and the public. Thus, the discipline is viewed as simple place name or location field on the one hand, or some other science on the other. For example, the work of a physical geographer is often taken to be geology or meteorology. Or, population geography is mistaken for demography (to which it may contribute its spatial analysis).

In sum, geography must be viewed as a field which touches upon almost every other discipline in that it deals with any and all phenomena from a spatial, regional and man-environment perspective. It is an art given that it can create beauty in its description that is hardly replicable, and it is a science in that it can and does utilize scientific precepts and is capable of predictions.

Willmott (USA): Geography is a user of methods and theories developed in cognate fields. In physical geography, for instance, we borrow regularities developed at smaller scales (e.g., from chemistry, physics, and biology) to see if they are useful, when combined with larger-scale geographic hypotheses, in explaining spatial processes and patterns. Newton's laws, for instance, are used in climatology to assist our analyses and explanations at larger geographic or spatial scales.

Geography similarly contributes to related disciplines by providing explanations for processes and phenomena which can be arrived at through analyses and models that are fundamentally location, distance, and space sensitive. Our position amongst the sciences and social sciences is akin to history's insofar as geography focuses on an underlying fundamental aspect of must natural and human processes — distance.

What does the geographical way of thinking contribute to the understanding of the world?

The set of responses regarding the uniqueness of geography's approach to understanding the world identifies three views. Clearly most respondents agree that the spatial or territorial approach to studying any human or physical phenomena is distinctly and uniquely geographic. A second and equally significant contribution identified by both the United States and Soviet groups is the study of man-environment interactions in their various forms, and viewed spatially, that is, analyzing man-environment processes as they vary from place to place at various scales. A third distinctive contribution by geographers is the regional approach, whereby geographers study the uniqueness of places and the value of grouping similar places into regions or regional systems. The latter is a particularly strong contribution of Soviet geography. Most geographers in both countries are acutely aware of all three uniquely valuable geographic approaches.

Brunn (USA): The uniqueness of geography derives from the way practitioners in the discipline look at the world, or at worlds, around us. Geographers are keenly interested in what goes on where, and why. The key word here is where. That query was, or is, just as important to the historical or prehistorical geographer as it is or will be to futuristic geographers. We are not satisfied only to know where something has occurred or will occur. We want to know why it is there and what it means (or the "so what"). That focus on the "whereness" of things can be related to places, locations, and environments. Geographers seek to know what an environment (physical, social, or economic) is like, what places are like (their meanings and values to people), and what locations are like (time and cost distances; social, political, and economic distances; interactions; and spatial organization). Space is a part of everyday life, and geographers attempt to understand and explain what part it plays in how and why we do what we do, where we do it, how we look at the world, and how we might better plan and organize environments and spaces. Geographers are also very much interested in holistic reasoning, thinking, and implementation. While it is not the only discipline interested in holism or interrelationships, our approach or way of looking at the world is unique because we look forward at spatial relationships or how things are distributed, organized, or arranged in space. The geographer's approach to holism, interrelationships, and integration spans the subject matter of the arts and sciences. What we continually strive to do is look for and discern order in what we see, map, measure, and know. Space at all scales (individual to international) is important to geographers because it is an essential part of the everyday world of all people and societies.

Pred (USA): The overarching and interlocking research traditions of human geography differ from those in other fields of human and social inquiry in that their primary focus and point of departure is what actually occurs on the ground, or concretely situated activities, landscapes, and expressions of spatial organization. Human geography also differs from other fields to the extent that its practitioners are sensitive to both the impact of humans upon the natural environment and the manner in which natural systems constrain and enable human activity.

Much of the work conducted by human geographers also differs from that of scholars in related fields in its concern for the contextual rather than the compositional. However, to suggest that human geography is "unique" on any of these scores would be pushing matters, for uniqueness requires fixed and impenetrable borders. Certainly there are anthropologists preoccupied with cultural ecology, historians of the Annales school, and others who begin with what actually occurs on the ground, who are sensitive to natural environment, and are concerned with the contextual rather than the compositional. And, just as pioneering advances in other disciplines may demand a penetration of what is "unique" to human geography, so does the making of conceptual and empirical advances in human geography often demand a penetration of what is "unique" to other fields.

Taaffe (USA): The three traditions in geography are the basis for distinguishing the geographer's approaches from those of other disciplines. Most world, national, and local-scale problems have significant spatial dimensions; synthesizing approaches heightens our awareness of world interdependence and provide an alternative to the undimensional approaches of economics, and so forth; and the relationship between man and the physical environment is becoming increasingly complex and is in need of study. Although other disciplines do work in related areas, geographers put the greatest emphasis on them, and only in geography do all three views figure prominently.

Akimenko (USSR): Geographical thinking influences the understanding of three facts: (1) natural and social phenomena are closely interconnected; (2) everything surrounding us is ordered in terms of territory; and (3) any local action results in global impacts, and global phenomena affect specific regions.

Annenkov (USSR): Specialists studying relief or soils, industries, or other sets of homogeneous objects in the geosphere are often apt to deny the existence of exclusively geographical thinking. Such specialists prefer to speak of the geographical approach to observed phenomena as merely the mapping of territorial differentiation of phenomena and its related conditions, as well as its ecology and the impact of environmental factors on it.

A century ago, the Council of the Russian Geographical Society in its decision "on the state of teaching geography in ... universities" proceeded from a different, integrational understanding of geography that studies "not separate phenomena but a group, or association, of them as well as inter-action laws." The foundation of the integrational geographical thinking was laid by A. Humboldt, E. Recluse, P. P. Semenov, and other outstanding geographers of the nineteenth century in various countries. Such thinking was based on the philosophical concept of universal interconnection and interdependence of phenomena.

N. N. Baransky explained the essence of the peculiar way a geographer thinks as follows: In 1938, in *Geography and School*, he said, "Geographical thinking is, first, thinking associated with territory, putting its reasoning on the map, and, second, it is connected and complex, not limited to the framework of one 'element' or 'sector', or, to put it differently, 'playing chords and not a single note'. " The Baransky school played "with chords" in studying the complex characteristics of countries and regions.

While earlier geographical thinking was discussed more often in terms of teaching geography at universities and schools, today its development also serves to solve economic, social, and ecological problems and to improve the methodology and methods of research over the geosphere.

Muller (USA): Geographers, unlike all other scientists, study the spatial dimensions of the human and natural world. They deliberately approach their scholarly domain in a broad way, examining the totality of landscapes and seeking to understand how various physical and artificial systems interact and combine to produce the myriad regional structures of the ecumene. Geographers also, like historians, study an entire range of human experience --- what goes on in terrestrial space and how the earth's surface is organized by humans and their cultures to achieve their goals. Through maps, geographers are able to study the vast complexity of the world, and their works contribute enormously to our knowledge of human behavior, human-environment interactions, and the more specific spatial relationships that have developed in the political, economic, social, and other spheres of systematic human and physical geography.

Tobler (USA): Geographers' approaches are not unique; the questions they ask are derived from theoretical points of view, as in all fields. In geography these theories are mostly based on asking why things are where they are or why they change from one location to another. But this is just a difference in emphasis.

Palm (USA): The geographer's approach to understanding the world may not be unique — that is, any individual geographer may be indistinguishable from his/her neighbor who is an atmospheric physicist, geomorphologist, historian, or sociologist. However, the field of geography is unique in that it is catholic in its interest in place --- not only admitting, but requiring the use of a wide variety of data sources and the integration of information about physical and human processes in order to understand the functioning of human-environmental systems.

de Souza (USA): Geography is unique because of its focus on the spatial dimension of life. In order to understand this dimension, the geographer uses a distinctive language, the language of maps. Geographers study places and their natural settings at the scales and levels of generalization at which maps, map symbols, and legends are necessary for understanding.

Demko (USA): Geography is unique in its understanding of the world in that it alone has the responsibility to analyze and describe any and all phenomena in relation to spatial and locational principles. The analogy of history conveys, at least in simplistic form, the unique role of geography. This is not to say that other sciences or arts do not, or may not, examine their object or phenomenon of interest from a locational-spatial perspective, but the responsibility, experience, tools, and philosophical basis for this contribution lie with geographers.

Myagkov (USSR): Geographical thinking introduces into the cognition of man the simple but very important idea of the great and harmonious diversity of nature. This concept lays the foundation for profound respect by all peoples living in "exotic" conditions. It is this concept that raises the question of natural limits of the applicability of even the most powerful technology and techniques born in industrially developed countries.

Solomatin (USSR): Geography gives society material to recognize itself as a part of earth's unique nature and its place and role in nature.

It also reveals the most rational ways to use nature. So far clarification has not been sufficient that the understanding of the physics of microcosm is neither more complicated nor more important than understanding the physics of the macrocosm of nature, particularly, the laws of physical geography. It is only geographical knowledge and progress that can improve this shortcoming of present-day thinking.

Mather (USA): The uniqueness of the geographer's approach is the focus on the reciprocal relations between the natural and human systems that constitute the earth. Our uniqueness may be thought of as a point of view, an understanding or an awareness of the connections among the physical, natural, biologic, human, economic, cultural, and historical, and a desire to employ integrative approaches.

Voronov (USSR): In answering this question, it is difficult to separate geography's contribution from that of other sciences in terms of the common perception. It does not seem quite right to speak about "geographical," "biological," "geological" thinking. I believe it would be more correct to distinguish materialistic and idealistic thinking as two mutually exclusive world outlooks and then to deal, not with thinking associated with specific sciences, but with the sum of facts and laws contributed by these sciences to our understanding of the world and affecting our cognition of the universe. The contribution of geography to the materialistic conception of the world is great. This contribution is determined by the geographical ideas about the material character of the world surrounding us and its development; the role of human society, which is determined by social laws; about man's impact on nature, often negative and irreversible, which man has to combat to preserve nature. At present, we have rejected the idea that nature exists for man and his needs only and that he can take from nature everything he needs in any quantity, altering nature at will. The existence of nature, independent of man, obeying its own laws that man must respect, is an important ideological statement which the world owes to geography. Finally, the concept of the diversity of the surrounding world, of the necessity of pondering actions directed toward changing this world in the interests of humanity is also an essential contribution of geography to the understanding of the universe.

Alayev (USSR): The main features of geographical thinking are "geoterritorial" complexity [dealing with sets of interrelated phenomena], concreteness, and globality. To express this brief definition as particular rules of behavior and approaches to research, the following may be helpful.

1. Mapping any phenomenon or problem immediately links it to the earth and distributes the elements in the same way they are

distributed in reality; the phenomenon can immediately be seen spatially, and many vague questions are clarified.

2. Studying a phenomenon in relation to its environment and other phenomena (everything is connected with everything) entails phenomena.

3. Every place on earth is unique, if only because of its geographical situation (from which other peculiarities result because of the zonality of the landscape mantle, the asymmetry of components, etc.). This feature should be considered both in research and applied geography.

4. The landscape mantle is unique. Changes made by man at one point affect, at different rates, not only neighboring regions but also large areas and even the entire mantle. One must think of the consequences of any and all actions.

Rodoman (USSR): The spatial diversity of the earth is the greatest economic and cultural heritage of man, which should be preserved and developed. Geographical thinking opposes standardization and extreme centralization in management. It urges that local conditions be taken into account and teaches us to value and respect the self-organization of humanity. The future should not be blindly constructed but should be developed based on knowledge of the objective laws of societal development.

Dmitrievsky (USSR): Geographical thinking means the ability to see the earth not only in its unity, implying a global-political perspective, but also in its internal diversity, in the territorial differentiation of everything that exists.

While the territorial differentiation of nature at the global level is clearly observed in the latitudinal and altitudinal zonality, the territorial differentiation of population and economy is a function of many variables, primarily the socioeconomic system of a particular state. The role of historical stages experienced by concentrations of population and economies in each country is also very important.

To understand our planet is not only to study the laws governing it but also to determine how these laws act under different conditions, which leads to the differentiation of nature and society. Only such an approach makes it possible to predict the development of nature and society, their interaction and interdependence.

It is impossible to separate space from time and thus chorological approaches from the chronological. The geographical approach envisages the analysis of spatial changes in nature and society over time. Without an historical approach it is impossible to discern any geographical laws.

Geography has analyzed these laws for centuries. The well-known geographical aphorism of Herodotus, "Egypt is the gift of the Nile," was one of the first examples of economic-ecological, geographical thinking. Geographical prediction attempts to optimize the processes of exploiting nature, and to provide intrastate, national, international, regional, and global monitoring.

Kagansky (USSR): A "sense of place" is characteristic of all people and groups (attachment to and familiarity with specific areas, landscapes), but geography rationally explains such senses. Geographical thinking focuses and transforms the rich knowledge of popular-geographical culture and introduces it into science (this is verified, for example, by geographical terminology which differentiates different phenomena and types of the natural environment (e.g., steppe, taiga, podzol, takir, and many others). Geography explains the ordered habitat of peoples and their real world.

CHAPTER TWO

Geographic Education

What types of geographical knowledge are needed for an educated person?

There is a clear consensus of opinion regarding the question of what geographical knowledge and skills an educated person needs. All agree that some factual base of locational information is required at a number of scales. Thus, one would be expected to have reasonable mental maps or spatial schemes regarding the location and distribution of major cities, mountain ranges, climates, political systems, and so forth. There is agreement also that such local, regional, and global data sets be understood as interacting systems in continual states of change. Second, the respondents argue that educated citizens should be aware of the importance of space and location and especially the need to make individual and societal decisions about location with knowledge and care about the impacts of such locational decisions. Finally, most agree that everyone should acquire at least elementary map reading and interpreting skills.

Demko (USA): An educated person should certainly be aware that geography is not merely the knowledge of where "things" are located. They should be aware that the location of things are only the beginning of geographical knowledge and that the distributions of all types of phenomena over space are dynamic, complex, and essential to understand. That is, rainfall patterns, distribution of missile sites, earthquakes, fertility, and so forth. are important because they affect each other, humankind, and they continually change and must be studied, modeled, and understood. Such a conceptual view of geography will diminish the importance of trivial locational knowledge and result in an educated person's curiosity about how important phenomena are distributed in regions of all scales, how these distributions are continually changing, and why it is important to support geographers and geographical analyses.

Morrill (USA): The secondary school graduate needs a basic understanding of the overall geographic structure of world societies— differences and similarities in environmental conditions, in livelihood systems, in social systems and in cultural forms. I believe this should be organized via such general principles as (1) the attempt of all peoples to adapt to and use their environments effectively; (2) the interplay of isolated development and uniqueness versus diffusion and internationalization; (3) the bases of conflict and competition for space and location, from the local to the global scale.

The college-educated person should achieve some understanding of the actual geographic spatial processes: settlement, location decisions, and conflicts, regionalization, migration, trade, international territorial conflict, and so forth.

Taaffe (USA): Rather than think of particular knowledge, I prefer the general goal of a greater awareness of interdependence among places as expressed in spatial interrelationships. This involves the habit of viewing phenomena on maps, looking for patterns, and comparing new patterns with preexisting patterns so as to detect relationships. Place location is a necessary basic vocabulary in this respect, but, in order to retain such knowledge, meaning must be ascribed to place. For example, what significance did the location or situation of such places as Belgium, Chicago, or Singapore have on their historical development?

Turner (USA): An "educated" person does not require advanced "geographic" knowledge (e.g., location theory, sedimentation rates). Rather, they need a basic understanding of the major elements of the biosphere and the "politico-econosphere" at a global level, including their comparative locations. They also need basic map interpretation skills, including the elementary means of spatial comparisons.

Akimenko (USSR): It may be easier to say what geographical knowledge man lacks today, rather than needs. Usually, people know their native area relatively well and the place where they live, although their knowledge is, more often than not, limited. For instance, an urbanite knows the territory beyond his city only as recreation places or picturesque landscapes, but is poorly acquainted with the character and way of life of the rural population. Conversely, for a country dweller, the city is a place for satisfying cultural and educational requirements, but it is not a specifically organized artificial environment where man interacts with the entire complex of sets of variables — people and nature, man and his habitat. Modern man lacks the ability to look at the territorial peculiarities of his own and others' ways of life through the eyes of other people. This is even more exaggerated in the case of viewing the world from the position of another social group, another class, or ethos.

Of course, this is not only a geographical problem but concerns a general world outlook of humanity. Nevertheless, it is desirable that man always ask the simple question — does his activity within an individual region, country, or world as a whole contradict the interests of other people?

Brunn (USA): An educated and intelligent person needs three kinds of geographical knowledge. First, some fundamental place-name knowledge of major physical and human features or location is required to be aware where things are going on within one's country, continent, and world. Some or much of that basic geographical knowledge should or could come from elementary classes in school. Second, with that rudimentary background the educated person needs to appreciate and understand the macro geographical patterns of the human or cultural and natural (or earth science) worlds. This knowledge includes basically a world regional awareness of major patterns of population, culture, religion, economies, political systems, climate, land forms, water, vegetation, and soils and the major reasons why these patterns are where they are. This global and regional knowledge is essential to be able to understand and to converse intelligently about national and world events and to raise legitimate and intelligent inquiries about Third World debts, environmental deterioration, nuclear winters, cultural conflicts, and superpower relations. The third knowledge set required is that of the inherent qualities of space and place, that is, the importance of space in understanding decisions that individuals, companies, societies, and governments make. Space is something that people do something in and with. How important are accessibility, interactions, networking, and time-space and cost-space convergence to those in marketing, transportation, communication, public service delivery, and site selection? Also how is space organized, reorganized, manipulated, and controlled? Who does it, how it is done, and what are the consequences? Behind the three areas of knowledge described above is an emphasis on maps and mapping. The educated person should not only see the value of placing materials on maps, but also learn how to interpret them and to have them give meaning to one's daily world and that of internationally oriented services and businesses.

Palm (USA): An educated person needs both a command of specific geographical "facts" and also an understanding of geographic principles. An educated person needs to know such facts as the nature and distribution of climates, natural resources, landforms, environmental hazards, culture realms, population concentrations, economic activities, and political-economic systems at the world, regional, national, and local scales. For example, the educated person should not only be able to locate political entities (e.g., nation-states) on a world outline map, but also be able to predict something about the nature of the environment and about human society at that place from his or her general stock of knowledge.

Among the geographic principles (as opposed to the "facts") that the educated person should know are those related to environmental modification (e.g., erosion and deposition, climatic change, and most importantly the impacts of human activity on such processes), and the spatial organization of society (the impacts of distance, direction and connectivity on location decisions, and the interconnectivity of societies in the world). Examples of such elemental, fundamental, and requisite principles are legion: for example, an educated person needs to know that a technological change in a given place, or an accident with major environmental impacts somewhere else, will have significant and predictable impacts in other places far from the area of origin.

O. A. Evteyev (USSR): In addition to purely applied geographical skills necessary in certain situations (the ability to get one's bearings by means of the compass, the sun, and local features, to use the barometer, maps) man must, first of all, be consciously convinced on the basis of the knowledge of geographical laws that it is necessary to maintain a strict balance in the relationship between man and nature based on the laws of the process of the interaction between man, society, and the environment. It is based on the study of these processes and laws that the main emphasis should be made in developing the knowledge and skills enabling man to competently form his relations with his habitat.

Man needs geographical knowledge and concepts of the diversity and, at the same time, uniqueness of the world and of the earth in its spatial, natural, and socioeconomic systems; of the interconnection and interdependence of the phenomena and processes of nature and society in their extent, structure, functioning, and development within the planet as a whole, its parts, regions, and places; of the main features and laws of the existence and development of territorial systems, natural and socioeconomic; of the methods of cognition of geographical systems, including, as primary, remote (air and space), and cartographic methods. It is necessary to know the main kinds, properties, and cognitive possibilities of air- and space-images and maps (general geographic and thematic) and to be able to use these materials to obtain geographical information; to get bearings and to make routes; to analyze and synthesize the theoretical peculiarities of phenomena and their systems; and to forecast the development of territorial systems. It is necessary to be able to carry out practical activity providing the efficient functioning of territorial systems, their development, and their use by society.

Rodoman (USSR): The educated person of today should have a more or less integrated visual conception of the earth, his country, his native locality as his mental map upon which he would imprint during his life span, geographic impressions. This image of the earth begins to form when learning geography at school and one consciously uses real maps and adds to them other information --- political, economic, ecological, and cultural-historical. Love of geography helps fill leisure with such interesting and multifaceted activities as tourism, a source of personal geographical discoveries, and a stimulator of ecological thinking.

Mather (USA): The educated person (a nongeographer) must, first of all, be aware of the world around them--where things are located, the relation of one feature or aspect to another, and the fact that the earth is a rational system, in general. Most elements of the physical environment, for instance, are located where one would expect, given a basic knowledge of the interrelationships among the various subsystems that make up the earth system. Humans are, at times, irrationally located (in areas of hazards, with little water, in areas of climatic extremes), but even humans cannot neglect the limitations placed on them by the natural subsystems. Thus, the educated person needs to be aware of the basic interrelations involved in the earth system and therefore understand how a modification of one subsystem can have an impact on another subsystem, either natural or human.

Muller (USA): A solid acquaintance with the major places in the world (countries, regions, physical features, large cities, and leading industrial areas) is absolutely essential to follow daily news events. In addition, educated people should have at least ten years of course work in regional and systematic geography between the third grade and the fourth university year. The British model is a good one; there are many others. I would stress that American and Soviet citizens take at least two years worth of geography courses about each other's countries.

Preobrazhensky (USSR): About two decades ago, many of my colleagues would have answered "to know the physical, economic, and political maps of the world, the nature and economy of one's own country." Today this is not enough. An educated man should be able to use various thematic maps and reference materials for everyday work and life. And, of course, he should be able to behave in accordance with the possibilities and limitations of nature.

The latter ability depends on the knowledge of the interconnections of the objects and parts of geographical reality surrounding us and with the knowledge of the possible reactions by nature to our activities.

However, Geography is not only an instrument of activity but is also an element of the culture of the age of the scientific and technological revolution. It is not just a literate and cultured person who should know that the fate of humanity and nature are one. We have only one earth and we, the inhabitants of the planet, are unique.

We must understand the unity of the landscape mantle, the interrelatedness of all its parts, geographical phenomena, and other interconnection with any kind of human activity, and with the spatially organized economies and settlements, and of the uniqueness of any corner of the earth. This understanding should heighten consciousness of our responsibility for the present and future state of the earth and people.

Meshechko and Omelyanchuk (USSR): The ecological situation today within individual regions and at the global scale can be successfully resolved if all humanity thinks globally and acts locally and regionally. But, to do this, each person should acquire a certain system of knowledge and skills.

How can the teaching of geography be improved in elementary and secondary schools?

Despite the fact that geography, as a distinct subject, is much more prevalent in the Soviet school system than in the United States, Soviet and American respondents reflect a strong degree of consensus in their answers to the question of improving the teaching of geography. The overwhelming recommendation from both groups is to remove the prevailing emphasis on teaching facts and to stress the teaching of geographical information via systems, spatial patterns, relations among systems, regions, man and environment, and similar dynamic concepts and methods. There is agreement also that maps should be more frequently and imaginatively used and that teachers must be better trained than at present. Also, there is agreement that professional geographers such as those at universities must interact more with teachers and advise and help school systems on geographic matters. United States geographers argue that geography should be taught in each grade in the United States as a separate subject and not as a part of history or social studies.

Demko (USA): Geographic teaching in the schools can and must be improved in the United States. All too often no geography is taught, or, it is combined in a social studies or world studies course. Clearly, geography must be taught as a separate subject at all, or at least most, levels of elementary and secondary school. Where geography is taught, all too often it is boring recitation of facts or memorization of place names. Unfortunately, too many teachers of geography are not well trained or trained at all in geography.

The remedies are many, expensive, and slow but must be applied. First, there must be more and better trained teachers of geography. Schools of education must work with research departments of geography to improve teacher training. Second, state school systems must be convinced to bring geography courses back to the curricula. Third, professional geographers and departments of geography must become more active in promoting and helping school systems, teachers, and so forth. Fourth, geographers must produce more interesting, exciting, current textbooks at all levels for the teaching of geography.

All of these recommendations are being initiated by a number of organizations and individuals in the United States, but the effort will be long, difficult, and frustrating. We must stay the course!

Cohen (USA): First and foremost, improvement can come by teaching geography (as well as history) as an independent subject and not as part of social studies. However, geography should not be taught as a rote, "content-oriented" discipline. It should focus on spatial patterns, spatial relations, and spatial systems, and especially on change.

de Souza (USA): The current status of geography in schools cannot be improved easily. Geographers often present an image of confusion about the nature and purpose of the discipline. To improve precollegiate geography instruction, geographers need to convince administrators and school boards that geography has a place in an already overcrowded curriculum. In order to do so, they must answer a number of questions. For example, what should the content be for geography courses in primary and secondary schools? How can the teaching provided through geography be strengthened? When the geography that students study is concerned with controversial issues, what teaching approaches are needed to ensure that students are made aware of the attitudes and values of those involved? Is enough attention given to the impact of political and economic processes and activities on geographical patterns and changes? How can the teaching of geography best be organized in both primary and secondary schools?

In recent years, answers to questions such as these have been provided by a few United States geographers involved in two related initiatives. The National Geographic Society, through its program of geography alliances, which are organized at the state level, is enhancing the position of geography and improving the quality of its instruction. The Geographic Education National Implementation Project (GENIP) is producing materials that are beginning to influence standards set by state education boards: textbook publishers are beginning to include auxiliary materials in geography; more and more elementary teachers are looking for ways to incorporate geography into the basic curriculum; the subject is more frequently the topic of inservice workshops; and teachers and curriculum developers are beginning to recognize that geography has a structure that goes well beyond simple place location.

Palm (USA): The teaching of geography can be improved in elementary and secondary schools in the United States in two ways. First, it is important that more courses of geography are required at all levels. The simple exposure of students in social studies classes to geographic inquiry, particularly in the senior high school experience, will have an impact of increasing the awareness of students of the mutual impacts of society and environment, as well as their knowledge of the human-societal and physical environment around them. These increased requirements can be brought about in several ways, including university entrance requirements and work with state legislatures and local school boards as curricula are reviewed and state requirements revised.

Second, we need better communication between university departments where geographic knowledge is being "created," and the public schools where such knowledge is disseminated. We need more programs of inservice training for teachers, better classroom materials, and generally better relationships between college or university geography departments and their local neighbors in the public schools. In the United States, there has been a good start at improving such relationships in recent initiatives by several professional associations, including the National Geographic Society, the Association of American Geographers, and the National Council for Geographic Education, but these efforts are minuscule compared to the activity needed to bring about the radical change necessary to deal with the current needs in the public schools.

Mather (USA): Elementary and secondary teaching of geography in the United States can be improved in at least two ways: (1) geography should be offered as a regular subject in each of the grades as is now the case in Canada and Europe; and (2) the teaching should provide greater stress on the uniquely geographical nature of the material offered. There is little chance that (1) the above will be widely introduced into American schools so that our greatest opportunity will be to concentrate on (2) above. It is clear that a considerable amount of what we would call geography is already being offered to students throughout junior and senior high, but it is not identified as geography. It also is not taught by teachers who have been trained in the geographical approach. Thus, students fail to grasp the unique geographical nature of the material and cannot see the interrelations among the natural and human subsystems. Improvement in teaching must be based on better geographically trained teacher --- teachers who are provided with course materials and texts that recognize geography for the integrative discipline it is.

Shevchenko (USSR): Presently, the state of Soviet secondary school geography as a subject addresses mainly educational and training tasks. Its role in the preparation of the younger generation for practical activity is still insignificant. The discrepancy between scientific geography and school geographical subjects is evident.

Geography at teacher training institutions is generally in the same situation.

Voronov (USSR): The difficulty of teaching geography at school in the USSR is that students are introduced to the main principles of geography too early in the curriculum when they do not yet know the fundamental material on which these principles are based.

Shkarban (USSR): The role of geography in revealing ecological problems was recognized in the new curricula for Soviet secondary schools developed in accordance with recent school reforms. Ecological education makes it imperative to demonstrate the complex and multiple system of interactions between nature and society. Such education organization makes it possible to show nature and its objects and phenomena, not only as a means of productive activity, but as its goal.

The implementation of the ecological principle in the process of teaching and education will make it possible to pass from the traditional natural science approach to the objects and phenomena of nature without considering the activity of man, to including man in connection with the reconstructing economic activity.

The ecological principle in teaching geography requires, not the introduction of additional concepts but the demonstration of new aspects and deeper insight into the existing concepts by revealing links and relationships. New terms and concepts are necessary as a form of knowledge generalization.

Bokov (USSR): An essential shortcoming of the curriculum regarding geography is the prevailing description of the geographical phenomenon at the level of large regions, tens, hundreds, and thousands of kilometers in area. Such large regions are understood mainly by means of cartographic and other models. It seems necessary to improve materials devoted to the analysis of objects, phenomena, and processes at the local level: small river valleys and river basins, slopes, soil and geological sections, microclimates, and microlandscapes. The phenomena of this spatial scale can be directly observed and studied during excursions.

It is necessary to introduce into schools elementary information concerning the most typical forms of interaction: positive and negative feed-backs, the circular form of interaction, and chain reactions.

School geography does not discuss concepts of cognition of the natural environment such as modeling and experimentation. Acquaintance with such dynamic techniques of studying natural systems strengthens the interests of school children in the subject.

Pancheshnikova (USSR): School geography must further strengthen its ties with psychology. One of the most important psychological principles formulated by A. N. Leontyev and S. L. Rubinstein, concerns the unity of consciousness and activity. In school geography this principle is revealed primarily in the strengthening of the double function of the curriculum and textbooks; they determine not only the content of geographical education but also the organization of the cognitive study of pupils and design methods for the development of their ability to learn. Pupils more often work independently in class; advanced teachers have begun to use such active methods and forms of teaching as group work, role playing, and seminars. Nevertheless, there is still a great deal to be done to implement this principle more consistently. Of fundamental importance for school geography is the attitude, worked out by L. S. Vigotsky, in which learning always precedes development. In practice this means that the optimal complex of learning should be provided at every stage of learning. Geography curricula, textbooks, and the processing of teaching from first to tenth form must not only expand subject matter (geological, geomorphological, climatologic, landscapes, economic regions, and other data) and respective skills but also increase the level of complexity of geographic issues.

The research results of a large group of psychologists under N. A. Menchinskaya and D. N. Bogoyavlensky devoted to the "mechanism" of the formation of knowledge and skills have been partially introduced into geography teaching. However, other concepts of teaching created by Soviet psychologists have not yet been used, including the theory of stage formation of mental actions created by P. Y. Galperin widely recognized in pedagogical science. It seems that the development of methodological solutions based on this theory would open additional possibilities to increase the efficiency of geography teaching.

Tobler (USA): Geography needs better teachers and a recognition that geography is not a study of place names. Students do not need to store information, they need to know how to find it, evaluate and judge it, and use it.

Muller (USA): Teaching of pre-college geography can and must be improved by retraining all teachers in the subject. Governments should fund, as a high priority, summer institutes and workshops to foster this. The National Geographic Society has made a fine start in this direction, and every state government and the federal government must be persuaded to continue funding on a permanent basis. The broadcast and print media should take a leading role in promoting geography.

Taaffe (USA): There should be more stress on thematic maps, particularly at the international scale. Geographers have been most effective in illustrating spatial concepts at the local scale, least effective at the global scale. More imaginative use of maps to examine changing world patterns of specialization, linkage, dominance, and urbanization are needed. The major need for the improvement of geography teaching, however, is quite simply more and better geographic training for teachers.

Brunn (USA): Geographers need to contact, consult with, and advise on a regular basis the appropriate government officials who work with science and earth science committees and advisers in schools. It is highly desirable to attend conferences of elementary teachers and to share with them what materials the professional geographic societies have available to assist them and to learn from the teachers how those of us in universities can assist them (maps, resource lists, visiting lectures). We need to offer credit courses for teachers being trained now and for those who have already graduated, but need additional courses for accreditation in geography. For current students, we need to encourage advisers in the colleges of education to have their students enroll in world regional, physical geography, and human geography courses. It is also advisable to develop an upper division course on Geography for Teachers or Teaching Geography. For those who have graduated, classes in World Regional Geography for Teachers, Physical Geography for Teachers, and Map Interpretation would be most useful. Courses must be geographic in content, not a general social or earth science course. University faculty must be very concerned not only with what is taught in the schools, but how it is taught. Efforts to promote geography must be done in concert with other geographers in the state, with school administrators, faculties in college of education, and teachers themselves.

Silenko (USSR): The new curriculum in geography introduced in the 1986-1987 academic year with its orientation to socialization, politization, economization, and ecologization should improve geography teaching and raise its importance among other sciences. The deletion of unnecessary details, the decrease of geographical nomenclature, and the reorientation to stress geographical concepts can only be welcomed. It is important, however, that school children realize the "epoch of great geographical discoveries" of the motherland and the earth. Memorizing excessively great amounts of dry figures and statistics should not be substituted for the joy of learning about the diversity of life.

At the same time, students must be taught how to find information on geography. Compositions on geographical themes might serve this purpose. The themes may be both at the global scale (e.g., the consequences of glacial melt in the Arctic ocean or the Antarctic, changes of the axis or orbit of the earth) and local scale (the drainage of the local bog, rational use of mountain slopes). It would be expedient to carry out competitions, ranging from local to international for the best school geography compositions and to publish them with specialists critically analyzing them. The formation of creative geographical thinking among people at large should become the most important task of the future. In addition to the shaping of geographical thinking, school geography might help to solve regional-information problems.

For instance, observations of the changes in animate life or erosion may be made. Schools, primarily in rural regions, might become a component of the monitoring service of the dynamics of the development of vegetation, soils, and other processes. Such works are expedient and profitable under the Hydrometeorological Service and research institutions. Some schools in Kirghizia and other republics (Sheko-Zakatal zone of Azerbaijan, Kharkov oblast, for example) have accumulated certain experiences in this direction.

On the whole, schools, as well as science, should stress geography as a positive and useful field.

Maksakovsky (USSR): It is, indeed, one of the "eternal" questions that cannot be solved once and for all. The world and society change, and the tasks, structure, and contents of school geography and methods of teaching also change. Therefore, in order to keep up, teaching should always be in the state of ongoing repair. Moreover, from time to time an overhaul is necessary. It is just such a period that Soviet school geography is experiencing now with the reform of the secondary school curricula according to which it will have 357 hours in the sixth to tenth grades. Very great changes have been introduced into the subject in the reform process.

To improve geography teaching at school is to implement the important didactic principle of the scientific aspect of teaching, to depict to the pupils a vivid and memorable geographical picture of the world --- to be exact, not just a picture, but a triptych including the whole triad: nature, population and economy. It is necessary to stress, at the same time, the organizing role of theory, fundamental knowledge with respect to facts, a different combination of courses of regional and local geography with general geography. It is laudable that a large group of geographers-scientists from the USSR Academy of Sciences have prepared the special handbook --- "The Elements of Constructive Geography" --- for school geography teachers.

Also, to improve geography teaching at school we must eliminate, or at least decrease, the existing disproportion between the principle of the scientific character of teaching and intelligibility, between the form of presentation and the content. It is no secret that dry, arid climate prevails on the pages of many school textbooks on geography, which absolutely diminishes interest. As one of the outstanding Soviet geographers, N. N. Baransky, wrote, the appeal to the minds of students should be complemented with the appeal to their emotions. It should be interesting to study geography! Only then will the lessons be swallowed with gusto. Indeed, the school textbook is the scenario not only of the training process, but of all the emotional spirit of the training process.

Finally, to improve geography teaching at school is to implement new progressive forms of teaching, the ideas of development, problem teaching, search for knowledge, ability to do independent work, elements of play and modeling, and to transfer the main stress from the reproductive to creative research skills by pupils.

It is natural that in the process of the school geography reform, Soviet scientists and specialists should themselves use in teaching methods, the rich traditions and great experience accumulated in our continuing experiment, primarily carried out within the framework of the Academy of Pedagogical Sciences of the USSR. Foreign experience, especially in the field of composing modern school textbooks, "playing at different scales" (from global to local), the study of an area from individual examples, the development of the active methods of obtaining knowledge, the use of audiovisual aids have also been widely used. It was quite natural, therefore, that the handbook of UNESCO *New Views on Geographical Education* published in 1986 in Russian, was sold out so quickly.

It should be noted here in this connection that the entire experience of contacts with geographers from western countries working in the field of education shows, unfortunately, that the vast Soviet literature in this field is little known to them and hardly used at all. Perhaps, it would be reasonable to prepare and translate into English selected Soviet works on the problems of school geography.

What are the main problems in improving geographical education for university students and the general public and what suggestions are appropriate for resolving them?

Suggestions for geographical education at collegiate and higher levels are many and varied among the Soviet and American geographers. Interestingly, both groups call for an improved public image of geography which reflects the rigor and importance of the discipline. There is general agreement that undergraduate programs and courses in geography must be enlarged and improved. The American respondents recognize the need to demonstrate to the public and other segments of society the applied and practical value of geographical research, which is much more appreciated in the USSR. The Soviet geographers emphasize the need to use the role of geoecological issues and environmental problems in changing curricula and attracting the attention of the public. Both groups call for better, more interesting, and accurate publications in geography both for public consumption and for textbooks.

Mather (USA): The main problem facing geographical education in the colleges in the United States is that we have so few students coming into college prepared (or even interested) in studying geography, because of inadequate high school programs. Thus, it takes a year or more to get students into our programs, and we can offer them geography for only two or three of their college years, often not in a logical format because of time constraints. At the same time, students are encouraged to look at subjects from narrow disciplinary viewpoints, and so they may not want to cross boundaries or to integrate materials taken from several disciplines. Geography could experience a real upsurge both in student interest and in the completion of significant research, if more geographers were willing to become involved in quantitative analysis, modeling, and utilizing information available from other disciplines.

Morrill (USA): Education would be improved by focusing on the geographic decisions that individuals, groups, and governments make, and why they make them. The public needs to understand that geography is not just travel and cultural uniqueness. The typical

53

American probably believes that the Soviet Union and its people are vastly more different from us than is actually the case. Better books and magazines by geographers would help. That is, what is life actually like in each others' countries. Who makes decisions about jobs and the layout of our cities, and what is the effect?

Demko (USA): A significant problem is that most university-level geographers are poorly trained in science; that is, they generally have weak backgrounds in natural science, mathematics, and computers (most were trained before the electronic revolution), and they tend to pass along antiquated methods and outlooks to students. Many geography faculty need to be retrained. Improve university faculty and our problems can be solved. If the problem goes untreated, university administrators will eventually apply their own solutions.

The problems geography faces at the upper educational levels are similar to those we face in other areas. Geography is perceived by educational administrators as unimportant or trivial or both because we have not been aggressive enough in our research and teaching to deal with relevant, real-world problems and demonstrate the value and complexity of geographic approaches. Too many faculty are dealing with issues that are insignificant from the public's point of view, and too many are writing and lecturing to each other in jargon. We also need to teach about current, important issues and write readable, good books for an intelligent, educated public.

Willmott (USA): Creating a more equitable balance between human and physical geography would also help considerably since, without physical geography, geography does not appear sufficiently distinct (to university administrators) from sociology, history, urban planning and so on. Our concern for the connections between the natural and human world is one of the strongest arguments for not dismembering our discipline. I fear many geographers have forgotten this.

Geographers should also publish important geographic works in the large scientific literature. Our nongeography colleagues' collective view of our discipline and its contribution to knowledge ultimately will be our salvation or demise. Talking only to ourselves (e.g., consider the within-department pressure in many universities to publish in "geography journals") tends to reduce our visibility, making us very vulnerable.

Cohen (USA): For university students, geography has to be taught as more than a set of techniques. The historical and philosophical bases of the discipline need to be treated in a serious, scholarly way. Analytic tools are a means to an end. The end is the study of overriding problems from a spatial point of view. The general public has to be shown how geography contributes to a better understanding of how place and space influence public policy as well as the more general course of human history.

Muller (USA): An effort to persuade universities to expand their geography programs should be launched. It will be difficult, but we should not shy away from it. In 1992 we have the Columbian anniversary and the IGU Meeting in Washington, D.C. We should develop as soon as possible a multifaceted strategy to enhance geography's exposure throughout the United States.

Dmitrievsky (USSR): Geographical thinking and geographical science itself have evolved together and have come through three main stages: descriptive, analytical, and constructive, each successive stage, not excluding but rather adding to the previous one.

It should be admitted that geographical education still errs with emphasis on rudimentary description. This circumstance makes itself felt not only in the development of geographical thinking but also in the prestige of the science itself and its role in society. And, the underdevelopment of geographical thinking is the bane of humanity in the system of man and nature relationships. The way to further the development of geographical education and dissemination of geographical thinking is rather diversified. First, geography should occupy its proper place in the curricula of secondary schools and be widely represented at all levels of education. The syllabi of courses should be primarily oriented to the development of geographical thinking based on the analytic-constructive approaches. Secondary school students should know the main physical and economic geographical laws and regularities.

Second, the curricula of all higher education institutions should contain geographical disciplines. Analysis of the problems of society's interaction with nature should be at the core of geography courses. Third, geographic thinking should be introduced in different forms (depending on the conditions of a particular country) to specialists of other disciplines. For example, there should be courses for specialists not familiar with constructive geography, problematic-geographical literature. Literature should be published especially for nongeographers and for the public by the mass media. It is no secret that many mass geographical publications in many countries attract the attention of readers, not with theories but with the description of exotic countries, peoples, and towns and by excellent photography. The same is true for television. A decisive change is necessary here.

An educated person of the contemporary world (and especially anyone with a higher education) should be able to use the scientific geographical approach. This is, therefore, a problem of "geographization" of all thinking and practice, which is important for the appropriate solution of most internal state and international problems.

Dubinsky and Riman (USSR): The way to further advance geographical education lies first of all in resolving the ecologization problem, and expanding the orientation toward the dissemination of geographical knowledge. The monograph *New Views on Geographical Education* (Moscow Progress, 1986, 463 p.) notes that the formation of geographical knowledge and skills is considered to be a complex and multilateral process in which no detail is unimportant.

Solomatin (USSR): The main means of popularizing geography and developing geographical education and thinking is to obtain successful practical results from fundamental geographical research. It is necessary to advocate, by all means, the positive value of geographical solutions to practical problems, geographical forecasting of natural environment evolution and its individual components and regions.

A sure way to improve geographical education is to establish closer contacts with relevant geographical organizations of the Academy of Sciences to involve leading specialists of these organizations in the teaching process and to use the technical base of these organizations for courses.

Taaffe (USA): One of the oldest problems faced by geographers at the university level is the necessity to make up deficiencies. If we were to try to teach modern geographic concepts to a beginning college student, it would be like trying to teach calculus to a student who knows nothing of basic algebra or even arithmetic. Geographic illiteracy, particularly as regards location, is widespread and has recently intensified to the point that it has been subject of several syndicated columns as well as commentary on TV network news. The trouble with the increased awareness of these locational deficiencies is that it leads to the impression that place-location geography is taught in the universities. It is not, but we are limited in how far we can carry the students conceptually by their woeful lack of a basic vocabulary in the form of a minimal locational framework.

Another problem within the universities is the tendency for geography courses and programs to be evaluated by administrators or other faculty who know little or nothing about the field. In many instances they have never had any geography courses in college, let alone courses in advanced spatial analysis, for example, and often have perceptions of the field based on a mixture of popular misconceptions and elementary school memories. The result is often a less prominent position in general education curricula for geography than is warranted by either the intellectual or societal goals of such curricula.

Within the discipline, the decline of regional courses poses a problem for college geography. In part, at least, this is associated with the failure to incorporate emerging spatial concepts into a regional course. In those few cases where it has been done the result is usually an interesting and challenging course which effectively displays all three of geography's traditions.

As for the general public, the main problem is a mixture of poor educational background in geography and the failure of the media, particularly local newspapers to provide coverage of international news. Even the commercial TV networks have replaced many of their "hard news" documentaries on international topics with programs stressing entertainment. The geographers themselves, of course, could help by producing more maps of relevance to current problems at world, national and local scales. In an increasingly visually oriented society, these are likely to be picked up by the media and form a basis for improving understanding of the problem in question.

Voronov (USSR): Animate nature is the most dynamic component of the biosphere. Little attention, however, is being paid to it in the curricula of higher geographical education in the USSR. When considering courses devoted to individual branches of geography, priority should be given to such topics as geological structure and relief. Established traditions in higher education geography are difficult to overcome. Changing the faces of all education to ecology, which we are all called to do, should make it possible for subjects devoted to nature and the interrelation of man and nature to occupy an important place. In addition to biogeography (with elements of ecology) there should be a course on "the ecological foundations of the rational use of natural resources and their conservation." It is not the number of such courses that matters but that all students are trained from the first to the last year. It is no less important to cut the descriptive material in regional courses and also bring general courses such as mathematics, physics, and chemistry closer to the problems of geography, which does not evoke great enthusiasm among instructors of the respective departments who give lectures and conduct practical studies in departments of geography.

Trofimov (USSR): The modern geographer-researcher must be able to pose problems, identify logical assumptions and concepts, develop research methods from geographical science; he must be able to solve the problems posed using contemporary technical means and, finally, to analyze the results obtained. The contemporary geographer should reject general speculations, attempts "just to add something" to the solution of global problems. He must be the coordinator in such solutions, the "supplier" of problems and leading executor, leader of complex research, and critical analyst when evaluating results. This model of geographer-researcher is implemented at Kazan University for training mathematically oriented economic geographers. For twenty years the curriculum has been developed with the methods of specialists using up-to-date means (including various types of computers).

Instilling geographical thinking presupposes using three ideas: geographical content, ecological approaches, and computers. The subject-matter cycle includes traditional courses, and the final course is Theoretical Problems of Geography.

The process of computerization starts with a course in computer modelling (first year) and ends with solving most complicated problems in a research paper. As the student thinks about working at his thesis beginning in the third year (at this time his future job is determined), the research project becomes not a formal event but of practical value to the institution, and it will affect his future work. The responsibility and quality of training increase and the expected annual economic benefit resulting from implementing student designs is estimated to be 200,000 to 300,000 rubles.

Palm (USA): The main problems in improving geographical education for university students are different from but not entirely unrelated to those of improving the geographic education of the general public. Within the university, the basic problem is the division of knowledge into subject-matter departments, and in this context, the absence of geography departments in some of our leading undergraduate institutions. Without a geography department, there is virtually no political support for the position of geography in the undergraduate curriculum and therefore no effort to add geography requirements or even electives. Since many of our future academic leaders and others in society who will have major voices in affecting university education are emerging from universities in which it was virtually impossible to take a university-level geography course, it becomes more and more difficult to reform higher education to reintroduce and even to maintain the place of geography in the curriculum. In order to reverse this situation before it is so serious as to be irremediable, it is important for individual geographers to take major risks to their own positions to plead the case for geography with key administrators.

In the case of the general public, geography is now known as either the dull course taken by many in the seventh grade, or as a popular field. Efforts to increase the numbers of significant books written by geographers and labeled as geography are important, as are continued public relations campaigns to increase the awareness of the general public of the very real contributions this field makes to human knowledge as well as to public-policy decision making.

de Souza (USA): People have a natural affinity for geography. They are drawn to geography, especially to its language of maps, because of their need to know where they are in both a literal and figurative sense. But, for many university students and members of the general public, geography has a chalk-dust image. Somehow professional geographers have made an exciting discipline dull.

The main problem with geographical education at the tertiary level is poor teaching. One way to redress this situation is to get the best minds back into the classroom. We must shift better teachers to prestigious institutions with lighter teaching loads and assign better teachers to advanced courses with small enrollments. Within departments, we must reward our most experienced teachers with less student-contact hours and place novice teachers in beginning courses where enrollments are large. It will take enlightened department leadership to encourage and reward teaching and to promote good teachers.

Geography is almost nonexistent as far as the general public is concerned. Lay persons have no idea what professional geographers do. Geography has no celebrities. Geography's public image crisis can only be solved by scholars who make people aware of our enterprise and how it contributes to a meaningful, civilized life on this planet. We need to produce the kind of scholarship and the kinds of good books that command attention.

Brunn (USA): The major problems geographers face in improving geographical education for university students and the general public are fivefold. First, many entering students to United States colleges and universities have very little, if any, formal geography background. Their last exposure to geography subject matter may have been in seventh or eighth grade when it was combined with other subjects in a social studies course. Some students may have been introduced to geographical concepts and terminology in high school courses in United States history or world civilizations, but these courses usually have a strong history content and are taught by historians.

Such a weak background in basic geographic concepts (scale, earth-sun relations, landscapes, environments, and regions), and in fundamental place-name geography means that college instructors have to spend some time in beginning geography courses in what might be called remedial geography. Second, much of what has been (and perhaps still is) taught in geography in the nation's high schools, colleges, and universities is poorly taught. The focus on minute details, memorization of often-termed useless information, and descriptive accounts of peoples and places leads to a misunderstanding of what much of contemporary geography is and what geographers do. The neglect of a focus on concepts, theories, and models, whether in examining cultural diffusion or landform formation or human spatial behavior, is lacking. Third, there are insufficient textbooks written at all levels in elementary, junior, high school, and high school by geographers for geography classes or for other texts with geographical content. Fourth, geographers need to confront and convince the university officials, university students, and general public that we address significant and timely themes and topics that are important in understanding the planet and that we teach these topics and concepts well. Fifth, the research we conduct must reflect some cohesiveness and integration and have meaning to students and those in the larger society.

The solutions to resolving the above dilemmas or problems are the following: (1) support efforts to provide pre-collegiate geography subject matter to elementary level teachers; this also means college and university geographers working with teachers, administrators, and teacher certification committees; (2) support all efforts to provide quality instruction in geography, including recognizing outstanding teachers, holding workshops, and offering retraining of teachers; (3) work with publishers and statewide textbook committees to improve the geography content of social studies, earth sciences, history, and world civilization books and ensure that geography concepts, terms, examples, and content are included in all curricula; (4) support efforts not only to train more teachers (especially historians) to teach geography, but to teach it better than it is now; and (5) endorse and publicize professional efforts to obtain agreement or a consensus on research priorities for disciplinary research, rather than encouraging and permitting anyone to do anything in the name of professional geography.

Turner (USA): Geography often survives well (large number of enrollments) where regional studies are an important element of the college curriculum, where the individual teachers are noted as superb, or where a program has demonstrated a logic to a series of courses with a perceived utility. The latter tends to be correlated to the perceived research contributions of the program at major universities.

Geography diminishes its contribution in higher education by presenting geography in only one manner, for example, spatial, physical. Each program should make it clear that multiple approaches are used in the discipline, even if that program does not offer some of them.

Geography cannot make much of an impact as a discipline on the lay public. Individual geographers can, however. The American public will accept geography and understand it better if it is reached by a major figure in the field.

Akimenko (USSR): Contemporary teaching is mainly the mastering of accumulated knowledge; the function of science is to acquire new knowledge. As a result, discrepancies arise between science and teaching. Students are introduced to past discoveries of science, particularly in the interface branches. This discrepancy is an objective and subjective one in dealing with nature and its use. There the scientist and the applied fieldworker "speak different languages." It is possible to eliminate this discrepancy between science, teaching, and application using the so-called "outstripping" method of teaching developed by Soviet scientist V. V. Milashevich. In "outstripping" teaching, both the "pupil" and the "teacher" are equal participants in the cognition process when the applied fieldworker teaches the researcher practical skills and the knack of "contact with nature," and, in turn, receives theoretical knowledge about the environment. Thus scientific-practical skills are acquired that focus on nature as the object of human activity and vice versa. Such teaching is very likely to be the required link in the reconstruction of geosocial activity and consciousness.

Zhekulin (USSR): Geographical education reflects (and should reflect in the future) the social order of society and the extent and direction of the development of geographical science. In the first stage of the development of the USSR (1917-1941), in connection with the acute need to take stock of natural resources and conditions of economic development, the geographical disciplines (geomorphology, climatology, hydrology, etc.) were just forming in science and education. The framework of geographical education in the country's universities was developed there. In the postwar years, these sciences and school subjects were enriched with new methods --- mathematical, geophysical, geochemical, satellite, and others. The sharp increase in major ecological problems over the last decade has brought about integrational tendencies in geography. This was seen in the formation of social ecology, geoecology, ameliorative geography, and the field of natural and economic regionalization. The "social" and the "ecological" came to be viewed as inseparable.

Geographic education also reflects this new tendency, although incompletely. The rigid structure of university curricula should be updated so as to retain and deepen geographical thinking based on the complex perception of both the landscape mantle as a whole and its various systems. The three main geographic paradigms --- spatial, geosystematic, and ecological --- should be reflected in geographical education.

The spatial paradigm (the most traditional) is expressed in the requirements for studying spatial relationships. It is reflected in numerous general earth science and regional disciplines. The geosystem paradigm focuses on the study of geographical structures as geosystems (natural, social, and integrated systems). The ecological paradigm reflects the ecological imperative; it is centered on man and the quality of the environment. Being essentially a metascience, it unites the other paradigms and is extremely important for geography.

In terms of the importance of ecological problems and the optimal use of the possibilities of geographical science, it is possible to outline the following recommendations for the curricula. First, incorporate ecology into geographical disciplines --- climatology, hydrology, social science, and so forth. Introducing ecology will widen the general educational functions of these disciplines and hence expand the range of the practical work of climatologists, hydrologists, soil scientists, and others. Second, ecologizing those geographical sciences which synthesize problems --- landscape science, bio-geocoenology, general earth science, and economic geography will also enhance their utility. Third, introducing integrated subjects --- social ecology, geo-ecology, ameliorative geography --- into the system of training can be implemented at universities and other types of higher-education institutions in different disciplines (meteorology, hydrology, etc.) and teacher training colleges can also be affected.

The dissemination of geographical thinking takes place through ecologization of engineering, agricultural science, economics, and other fields. Because of the geoecological approach, scholars in "narrow" disciplines will find it necessary to consider such issues.

CHAPTER THREE

The Relevance of Geography

What useful knowledge and advice
do geographers provide to
business, government, and other institutions?

Soviet and American geographers provide a variety of responses with examples in answering the question regarding useful knowledge and advice geographers provide to institutions. The United States' responses reflect an emphasis on business and the private sector and a grave concern for the lack of public knowledge about geographic applications. The Soviet geographers, however, describe a large and fascinating array of examples of public sector applications in many specific projects and developments in the USSR. The entire set of responses provide an excellent view of the actual and potential application of geographical expertise to all sectors of society.

Vorobyov (USSR): Geographical sciences in the USSR have always been of great importance because of the country's size and the diversity of conditions. Geographical knowledge is useful to the national economy because it reveals the role of territorial factors in the country's economic and social development, shows the territorial differentiation in processes and structures for planning and economic development, and stimulates wider interactions among economic regions and republics.

Geographers provide information for economic development of both the country as a whole and individual regions and economic branches. Geographers are directly involved in planning and design development and are not limited to merely transferring information. Such work, both on individual components of the natural environment (land evaluation, climate, vegetation, water resources, etc.) and the total natural complex is of particular importance for developing regions and estimating their resource potential.

Of great importance also are plans, developed by geographers over the last few years, for rational utilization of the environment, and the conservation and improvement of the resource base. For example, the Institute of Geography, USSR Academy of Sciences, has published two valuable atlases: *The Natural Environment and the Natural Resources of the World* and *Snow and Ice Resources of the World*.

Armenian and Siberian geographers have come out with useful and practical recommendations for the preservation of the unique natural complexes of Lake Sevan and Lake Baikal. Siberian and Far Eastern geographers actively participated in the compilation of a territorial complex scheme for environmental conservation in the Baikal-Amur zone. Geographers of the Ukraine, Byelorussia and Lithuania have prepared practical recommendations for rational utilization of the environment in their republics.

Recommendations from individual subfields of geography are also important. For instance, geomorphologists of Moscow, Moldavia and Kazakhstan have prepared materials on dealing with natural catastrophes (mud streams, land slides), and glaciologists of Moscow, Uzbekistan and Kazakhstan have studied the formation of glacial run-off, which is very important for Central Asian rivers. Siberian biogeographers have studied and mapped vegetation zones along the BAM railroad line and estimated fodder resources of intermontane basins.

Geographers have prepared recommendations for rational deployment and development of productive forces. Suggestions by Siberian geographers on territorial-productive complexes, population issues and rational settlement processes are noteworthy. The staff of the Pacific Ocean Institute of Geography have suggested a network of discrete economic regions for the Baikal-Amur Railway zone. Geographers in a number of republics have suggested ways to rationalize migration processes to utilize labor resources.

Recommendations of geographers are considered either before the start of national economic development projects, at the pre-planning stage, while developing pre-planning documents (such as regional planning schemes, schemes for deploying and developing productive forces), or during the inspection of national economic projects. Geographic advice is also considered when extreme ecological situations are being analyzed.

Examples include the work done in compiling the regional planning scheme for Irkutsk region, the territorial complex scheme for environmental preservation in the BAM zone and Lake Baikal basin.

Geographers working in the Altai area organized inspections of the Katun hydropower plant project via a scientific conference on the "Geographical Problems of the Katun Basin in Connection with Energy Development." After discussing the project in detail, the participants pointed out the inadequacy of environmental protection measures and worked out a concrete projection of direct and indirect impact of the hydrotechnological project on the environment. Questions were raised about complex development of the Katun basin in connection with energy development, the need to take into account the peculiarities of the environment, and the necessity of minimizing the harm done to nature as a result of economic development.

Geographers' recommendations are very important when improving extreme situations created in individual regions. For example, Lake Sevan and Lake Ladoga can be cited (where scientists of the Institute for Limnological Studies of the Soviet Academy of Sciences gave valuable recommendations). The problem of the Baikal was aggravated as a result of construction of pulp and paper industries in the lake basin. The recommendations of scientists of the Siberian Department of the USSR Academy of Sciences (the Limnological Institute, the Institute of Geography, the Institute of Forest and Wood) were taken into consideration in the Resolution by the Party and the government, which was directed to improving the exploitation of the natural resources of Lake Baikal and protecting its unique natural properties.

Prokhorov (USSR): As far as geographical knowledge is concerned, the picture is clear. From childhood, we learn from geography about the peculiarities of nature, economic activities, and the way of life of the population in a region, country, neighboring countries, and the whole world. Contemporary man is not fully understood without this general cultural background. But, in addition to this, geography has another very important role. It provides practical guidance to society to help in making decisions in the field of economic management, improving living conditions of the population, modernizing the construction and reconstruction of industrial plants, and so forth. Such examples are numerous. I shall present a number of examples from my specialty, medical geography.

The scientific activity of medical geographers is closely connected with solving many vital economic and social problems. For example, the Ministry of Public Health of the USSR was developing a long-term scheme for the distribution of public health services. Medical geographers contributed a set of maps showing the peculiarities of local conditions and the specific problems in providing medical aid to people in different regions of our country. One map illustrates situations in which a district physician is responsible for several apartment buildings, located near an outpatient clinic, and with almost a thousand more people living in each.

Medical service must be organized quite differently in a region where a thousand people are scattered in three to four villages separated many hundreds of kilometers and without suitable roads. More than twenty economic development projects for regions in Siberia and Far North and related settlements were carried out using medical geographical works and recommendations.

Information, prepared prior to the onset of development, concerning hazards such as diseases of a local nature, the presence of blood-sucking insects and ticks, poisonous plants and animals, and so forth, makes it possible to prevent illness and accidents among the workers of the new regions. Such works were prepared for the Baikal-Amur Railway zone, the West Siberian oil and gas complex, the Kansk-Achinsk basin and other major projects.

The concern of medical geographers is not limited to initial development projects. Unreasonable use of natural complexes, construction of large enterprises ignoring the peculiarities of the natural environment, poorly designed towns and settlements that are unsuited to the region --- all can have negative impacts on the way of life of the people and their health. Therefore, medical geographers participate, with designers, in working out project scenarios at all the stages of development, from the first tents to the completion of the construction of large towns and industrial plants. The main criterion is the creation of the most favorable conditions for work, life, recreation, and the good health of the people.

Another example relates to builders choosing a site for a future town in the BAM zone. The conditions in the region are complex --- severe permafrost, high seismicity, and swamps. As a result of helicopter flights, the use of land-rovers, and field trips on foot, a suitable settlement site seemed to have been chosen in terms of local standards. The specialists approved this version, but the medical geographers objected because it was necessary to fell the only large forest in the region, leaving citizens of the future town only swampy wastes for recreation, sports, and camping. As a result, an alternative was adopted that used the materials left over from building the mining operation to cover the swampy sites, leaving the forest as a suitable recreation area.

On the initiative of medical geographers, the general plan of the town of Ust-Ilimsk in Siberia was revised. The first version would have placed the town in dangerous proximity to an industrial zone and the unfrozen polynya of the dam. This polynya, extending over many kilometers, is the source of permanent, intensive fog during the long Siberian winter. Medico-geographical maps, compiled by medical geographers, of the distribution of diseases characteristic of some natural localities were also of great importance. Maps of the preconditions of diseases due to the excess or deficiency of different elements in water, soils, agricultural crops, and food in certain regions were also used. From these maps, regions are identified where people should be vaccinated and microelements added to their food (e.g., fluorine for water or iodine in common salt).

Myagkov (USSR): Engineering-geographical recommendations contribute to preventing socioeconomic damage and reduce the cost of measures aimed at preventing damage from natural disasters. Economic losses resulting from floods, hurricanes, earthquakes, volcanic eruptions, and so forth, in populated areas are astronomical.

The impact of natural calamities increases as civilization evolves. The causes, for example, include: (1) new development in less hospitable territories; (2) more expensive mechanisms and more sophisticated productive links between different enterprises; which makes every accident more costly; (3) new technologies and new types of construction and devices which are more vulnerable to natural phenomena; (4) local anthropogenic conditions and appearance of new types of calamities in mining and other industries; (5) future changes in the distribution and force of many kinds of natural calamities resulting from climatic changes; and (6) the disregard of natural hazards on the part of those who are enchanted with the growing power of technology.

Levadnyuk (USSR): In Moldavia, a serious problem existed in connection with preventing landslide damage to the support structure of high-voltage power lines. The Moldavian Energy Administration addressed this issue in cooperation with the Academy of Sciences of USSR and specialists from the Laboratory of Exogenous Geomorphological Processes in the Geography Department. These contacts made it possible to work out a complex method for estimating geomorphological conditions when selecting power-line sites in regions of landslide risk in Moldavia. This approach, known as engineering-geomorphological analysis, has proved successful in designing twenty electrotransmitting lines. They have all been built, are in good technical condition, and are functioning normally even during periods of landslide activity in the republic.

Engineering-geomorphological analysis includes principles and criteria for rational deployment of power lines in the landslide conditions of Moldavia. It is recognized by the designers of the republic and, currently, no electrotransmitting line with a capacity of 35 kilowatts or more is designed without employing it. The practical results of engineering-geomorphological analysis are the basis for *Recommendations for the Rational Distribution of Electrotransmitting Lines in Landslide Risk Regions (The Example of Moldavia)*. These recommendations have been approved by the Technical Council of the Moldavian Energy Ministry and are recommended for use in the republic. Their application makes it possible to reduce costs of research considerably, to increase design quality, and to assure reliable exploitation of the power systems and protection of valuable lands.

Bukharitsin (USSR): The Department of Hydrometeorology of the Astrakhan Zonal Hydrometeorology Observatory has issued a monograph *Dangerous and Especially Dangerous (natural) Hydrological Phenomena in the North Caspian.* All the data and observations of many years at all the hydrometeorological stations and services of the North Caspian were generalized and used in the work as well as ice air-survey data, satellite information, data from regions of natural calamities, and information from observers and witnesses and literary sources. Such work has been done for the first time in the Caspian area.

The results and conclusions obtained are of great practical importance. They are the basis for consultations and recommendations used by many research and design institutions and industrial associations and organizations, which allow them to consider the peculiarities of the hydrological regime in their activities, to predict possible harmful effects from hydrological phenomena, and to take necessary measures to remove or minimize them.

Peshkov (USSR): The experience of the Georgian Coastal Protection Agency, the first research-productive association in the USSR created to protect seashores from erosion, is an interesting example of using coastal zone expertise. The intensification of economic activities on seashores over the last few years has brought about unfavorable ecological implications. Shore erosion processes are known to take place all over the world. About 25 percent of the coastline in the United States is subject to severe erosion, at a rate of 4 to 5 meters a year in some places. Similar rates of shore retreat are observed on the Mediterranean coast of France. The sea annually washes away large areas of beaches in Japan. After the port of Madras was built, the Lee coast retreated almost 1 kilometer over ninety years. On the east coast of the Caspian, waves have washed away a strip up to 2 to 4 kilometers since the end of the last century. In the region of the town of Poty on the Black Sea, up to 1 kilometer of shore loss occurred over forty years.

Such seacoast damage was done not only by the ruthless extraction of gravels and sand for construction purposes but also by various technical buildings, even those specifically created to protect the coasts from erosion. Many designers of these protective structures --- wave reflective walls, underwater breakwaters, and so forth --- calculated the structure's resistance to sea impact and did not include the effect that these structures have on the dynamics of drifts and other processes in the littoral zone. The character of the coast, the origin of drifts, the direction of their transfer, and other natural factors were ignored. As a result, the expensive structures were rather quickly destroyed and did not reliably protect the beaches. They were replaced by new, more massive ones which violated the natural regime of littoral processes even more. The erosion extended to considerable areas of the coasts.

A radically new direction in coastal protection was implemented in the 1980s on the basis of complex analyses of littoral zones. The Georgian Coastal Protection Agency utilizes new, scientifically designed techniques. Protection is based on removing part of the debris in areas of drift-flow formation, resulting in the shift of the drift along the coast bringing about the artificial formation and accumulation of beaches. For example, on the coast from Gagra to Pitsunda (30 km) in 1982--1985, extensive areas of debris were removed, resulting in an area of beaches exceeding 20 hectares; in some places, beach width reached 60 to 65 meters. The method of the artificial formation of beaches in that area is twice as cheap as the proposed construction of breakwaters, and so forth.

Thus, the constructive geography of the littoral zone already yields more economical and ecologically reliable solutions than purely technocratic approaches to the protection of coasts.

Litovka (USSR): The Institute of Socioeconomic Problems of the Soviet Academy of Sciences has carried out a number of research works for the purpose of planning and management, with geographers playing leading roles.

1. The notion of a qualitatively new planning document, the Unique General Plan of the Development of Leningrad and Leningrad Oblast (a comprehensive socioeconomic entity) has been developed. Thus, the obstacles to applying common measures over the territories of the town and the oblast at different administrative levels are removed. Research is being conducted to reveal regional settlement systems at oblast and lower levels as well as to determine the existing functional boundaries of towns and rural settlements that are the basic elements of regional systems. Based on such research, proposals for the improvement of the administrative-territorial arrangements will be prepared.

2. Geographers of the Institute of the Socioeconomic Problems of the USSR Academy of Sciences, while preparing the State Complex Programme of Research and Use of the World Ocean in the Interests of Science and the Economy, pay serious attention to the study of specific economic structures forming at the interface

of ocean and land. This is most vividly manifested in the form of maritime socioeconomic complexes. The ever-increasing role of such complexes requires an efficient policy of operational and strategic management that cannot be achieved without the scientific substantiation of the boundaries of maritime complexes and their place in the state system of economic and social regionalization.

The same program includes the White Sea project in which geographers of the Institute of Socioeconomic Problems investigate problems of increasing the biological productivity of the White Sea. In particular, they have developed methods for the socioeconomic substantiation of a complex of economic measures on measures to develop a maritime economy and to estimate its the settlement system forming on the White Sea coast. Methods of developing the socioeconomic resources of Lake Ladoga and its basin are also being developed. Their importance is obvious, given the unfavorable ecological situation forming in the basin.

3. Research work carried out on the optimal use of natural resources of the Leningrad socioeconomic complex are also of applied significance.

Trofimov and Khuzeyev (USSR): Economic geographers of Kazan University have developed a unique system of automatic regionalization and have applied it to the agro-ecological and agricultural regionalization of the Tatar Autonomous Soviet Socialist Republic (ASSR). The results obtained were very useful for planning the development and distribution of agricultural production in the republic as well as for designing urban development schemes for Kazan, Astrakhan, and other cities, taking into consideration such unusual variables as the historical, architectural, and artistic values of buildings.

Another methodological approach that includes elements of simulation modeling and fuzzy sets was used when identifying optimal routes for passenger transport in a number of towns of the Tatar ASSR. Such schemes, which minimize the time taken by transport travel, have been introduced in the towns of Naberezhnye Chelny, Nizhnekamsk, Almetyevsk, Bugulma, and Chistopol. Currently, intensively developed geographic methods using the theory of interest coordination (making compromise decisions) have solved a number of interesting practical problems. Using these methods, the optimal scheme for bus routes for the town of Chistopol was determined, yielding a better solution than simulation methods. The degree of rationality of the territorial service industry was estimated and optimal schemes for the distribution of service enterprises in a number of administrative regions of the Tatar ASSR were determined. These important works were performed in direct response to requests from interested organizations.

We consider the methods of the theory of compromise decisions to be most promising for solving geographical problems, since practically any geographical situation can be conceptualized as a conflict between parties with noncoinciding interests and where a solution presupposes finding a certain compromise that takes into account the interests of all sides.

We consider the theory and methods for the analysis of the structure of geographical complexes for modeling and managing purposes an important contribution of geographical science. The use of such characteristics of the structure as its diversity, efficiency, stress level, and so forth, make it possible to study geographical complexes in terms of controlling their functions and development. The Nizhnekamsk territorial productive complex became the object of such an analysis.

We believe the above examples illustrate convincingly the possibilities of modern geography to solve practical problems when one applies a unique conceptual framework, and well-developed and interconnected methods and techniques for realizing mathematical-geographical models.

Vikulov (USSR): From ancient times to the present, geography and all the derivatives of our science have given and are giving an infinite quantity and diversity of knowledge for practical application. This knowledge is copious as far as the territorial-spatial dimension is concerned, ranging from the local scale to vast regions of the earth's surface, the oceans, and even to the far cosmos.

I will single out only one problem from this colossal range of possibilities, namely, the problem of Lake Baikal, the region where I have spent all my life.

The rational use of nature in the Baikal basin presupposes the conservation of Lake Baikal and, consequently, means a wise approach to the distribution of productive forces here. Enterprises characterized by particularly poisonous emissions or having no reliably efficient technical techniques for purification are not desirable in the Baikal region. Moreover, their deployment on the territory of the protected area of the lake was forbidden as early as 1969 by a special resolution of the Soviet government.

In spite of the ban, some ministries have constantly tried to deploy some new enterprises in the region, including those polluting the natural environment. In this situation, the role of geographers and other scientists capable of heading the ecological inspection of the technological designs of industrial enterprises sharply increases.

For six years geologists, geographers, and representatives of other sciences conducted ecological inspections of the Trans-Baikal appatite plant. The plant was supposed to be situated on the Selenga River not far from the town of Ulan-Ude, the capital of the Buryat Autonomous Soviet Socialist Republic (ASSR), at a short distance from Baikal. Geographers showed that all the technological functions of the plant, from the extraction of phosphorus-containing ore to the storage of the waste products were associated with negative consequences for the natural environment in general and Lake Baikal in particular. The authors of the plant design made attempts to develop a number of environmental protective measures. Repeated ecological inspections, however, persistently proved the measures to be inefficient. It became clear by 1986 that there are no reliable technical or technological solutions at present that might prevent damaging the environment and the particular ecological conditions of the waters of Lake Baikal. The construction of the plant was stopped.

I would like to emphasize that termination of construction was not in itself the purpose of our inspections. This decision was made in the course of ecological examination, when it was proved unambiguously that there were no ways to prevent the degradation of the ecological system of Baikal while the Trans-Baikal appatite plant operated. It goes without saying that closures and bans are not the goal of geography. Thus, I submit another example to illustrate the geographers' approach to the problem of deploying new enterprises.

The Buryat ASSR urgently needed a hydrolysis-yeast industry to meet the needs of cattle breeding. The relevant organizations suggested constructing the plant on an industrial site in cooperation with the Selengan cellulose-cardboard combine. However, the hydrolysis industry is characterized by especially harmful flows (when operating traditionally), and the site suggested was situated only 40 kilometers from Lake Baikal. Having carefully studied all the questions connected with hydrolysis production, geographers and biologists recommended against the Selengan site and any other within the drainage area of the Baikal basin. Instead, six sites in the territory of the republic were suggested, all situated beyond the water protection zone. It was proved that two of them are most suitable for the construction of a hydrolysis plant.

The practice of ecological inspection of enterprises in the Baikal basin has shown, however, that positive results are achieved only when geographers unite with biologists, ecologists, economists, geologists, and so forth. It is in such cases that the necessary complexity and scientific substantiation of recommendations are achieved.

Alayev (USSR): I would prefer to formulate the question as follows — in what cases is advice given by geographers more competent than advice given by representatives of other sciences?

At the national level, the power of geographers (their methodological approach, the complexity of their knowledge) consists of their ability, on the one hand, to evaluate more completely locational decisions (at least from the point of view of geographical situations and resource provision); on the other hand, it means to predict, based on the knowledge of the laws of evolution, dynamics and functioning of the landscape mantle, the performance of the natural-economic system. Finally, geographers are recognized specialists in the field of spatial classification, that is, regionalization, zoning, and, in general, taxonomy. Therefore, their knowledge is at a high competence level when addressing the following questions:

1. Estimating the best locations in a region or a part of an area in terms of its potential.
2. Forecasting (geographical prognosis).
3. Regionalization and, in general, zonation of some actions, measures, policies, and so forth.

At the global level, geographers must take part in the solution of current global problems. These problems, difficult in themselves, are insoluble without geographers' participation. Geographers do not yet fully understand the main laws of harmony between nature and man, but they possess all the means (accumulated knowledge, methods, and methodology) to be able to discover these laws.

Brunn (USA): There are a large number of professional geographers who work in the public (government) and private (business and industry) sectors. They are sought and valued for their ability to handle large amounts of geographic data that deal with locations and areas. They are needed for their expertise in mapping and information systems; their experience and understanding of a region (United States or foreign area) and regional problems; their understanding of spatial systems (economic, energy, demographic, etc.); their abilities to integrate concepts and materials from the human and physical sciences in planning (land use, regional, urban); their regional development expertise and hazard perception (natural and technological); as well as their experience with policy and mitigation. Professional geographers are able to provide useful knowledge and advice in the technical areas, including cartographic; computer cartographic; aerial photography;

remote sensing; geographic information systems; regional integration efforts; spatial organization; human-environment problems, which are the focus of public policy; spatial strategy; decision making; and spatial statistics.

de Souza (USA): As members of one of the oldest disciplines, geographers have always provided valuable information and advice to their important constituencies. For example, from the fifteenth to the nineteenth century geographers were on the voyages of discovery and their reports back home were the "big" news of the day. They helped bring to the consciousness of European business and government leaders the first coherent pictures of the broad environmental patterns of the earth, and information about the cultural diversity and natural curiosities found in other lands. Geographical societies with their exploration and publishing activities were seen as essential to European achievements overseas.

Today the sorts of knowledge and advice geographers provide fit under three general headings: human-environment, spatial organization, and inventorying and monitoring research. Under the human-environment theme, geographers provide information about the best match between the environment and the "product" realized from the environment. Under the spatial organization theme, geographers provide answers to questions about physical and human structures. Inventorying means collecting and analyzing information especially for planning use of human and natural resources.

Monitoring research, which is a blend of the theoretical and the practical, is intended to provide geography's constituencies with information about change and to find out whether changes are harmful to people, the resource base, or both. The geographer's constituency is one of the great challenges of applied geography today. Geographers are well suited by training and viewpoint to help meet this challenge. Nonetheless, our impact on our important constituencies is not as great as it should be.

Taaffe (USA): Locational analysts, cartographic analysts, environmental impact specialists, transportation planners, and others work for business and government. The State Department, Department of the Interior, state development agencies, and city and regional planners are among the government employers of geographers. Among the private employers are retail chains, banks, and consulting firms. Geographers bring knowledge and advice concerning locational and environmental problems faced by these organizations. More fundamentally, geographers call attention to: the interrelatedness of phenomena and the interrelatedness of places; the significance of the spatial dimensions of the particular problem at hand; and the shortcomings of considering only the economic or only the political aspects of a problem.

Palm (USA): Geographers provide information and advice to business and government in many contexts. Geographers are actually employed by businesses and government agencies to "do geography," that is, to analyze patterns of economic activity and population dynamics along with environmental impacts, in order that business and government can better perform their own functions. University-based geographers may also work on a contract basis with such agencies to consult with them on particular empirical problems. Four actual examples are:

1. Geographers knowledgeable in the problems and practicabilities of risk communication working on a contract basis with the U.S. Geological Survey to help them prepare their materials concerning earthquake hazards into a form more easily understood by the general public;
2. Geographers knowledgeable in economic geography, particularly that subfield known as retailing geography, working for such multi-locational corporations as Dayton-Hudson company to analyze population dynamics, traffic patterns, economic forecasts, and urban growth patterns in order to select optimal locations for new store complexes;
3. Geographers working for ski resorts to do that portion of environmental impact statements dealing with long-term air pollution associated with resort development;
4. Geographers knowledgeable in survey research working for such federal and state agencies as Bureau of the Census in order to refine methods of obtaining and analyzing large-scale survey information.

Muller (USA): American geographers provide an enormous range of knowledge and advice to the public and private sectors. For government, geographers are heavily involved in planning at every level, for land-uses, resource development, population projections, and in countless other pursuits. Political geographers help many agencies understand today's changing world in the areas of boundary change, terrorism, refugee flows, and in keeping the federal government up-to-date on geographical information throughout the world. The U.S. Geological Survey is involved with mapping programs of every kind; economic geographers advise on the changing location of agricultural and industrial patterns; urban geographers consult on housing, transportation, and policy - development programs of all kinds.

Research programs of the National Academy of Sciences and National Science Foundation are also influenced by geographers. Many of the same activities go on in the business world, where geographers play a major role in business location decisions.

Willmott (USA): Geographers provide advice to industry about where to locate, to government on boundary disputes. They provide a wide range of information to many on cultural practices that favor or preclude the sale of certain products, available solar energy, air pollution sources and concentrations, weather patterns and health, leaching of toxic materials through land fills, climate information for travel agents, advice on the impacts of land use, market-region descriptions analysis, and many other areas.

Morrill (USA): Knowledge and advice of geographers is given to business and government. I've devoted much of my professional life to this endeavor. Essentially, we need to inform about the options for and the consequences of environmental and locational decisions. All individuals, groups, enterprises, and governments make countless location and movement decisions. If we understand these processes, we should be of great assistance. (Often our knowledge is inadequate and we substitute what we would like to see, thus reinforcing poor decisions).

Demko (USA): There are a large number and wide array of examples of applying geographic expertise to the public and private sectors in the United States but, much of this work is not well known. Many geographers work in state and local government, business, and the federal government, but usually their title or designation is not that of geographer and their work is not identified as geographic. Our profession needs to be more active and creative in identifying these geographers and encouraging them to make their work known. There are some excellent examples of rigorous geographic work being done by colleagues at the National Aeronautic and Space Administration (NASA), in state government, and in private development firms. These contributions must become more visible for our young geographers and the public.

How widely applied are the recommendations and suggestions of geographers?

Regarding the question of how widely applied the recommendations and suggestions of geographers are, there is an interesting coincidence of opinions among Soviet and American geographers. They agree that geographers' input is not widely enough sought nor their suggestions heeded in either society, and, to their detriment. A number of Soviet geographers argue that geographic advice is accepted after long delays and is thus too late to prevent problems. Both groups provide a few examples of success but note that they are the exception rather than the rule.

Brunn (USA): Geographers offer suggestions and recommendations to many businesses, industries, and local, state, and national governments. My belief is that geographers provide a good amount of advice, especially considering the relatively small number of professional geographers in the United States. I know geographers who are sought for their expertise in advertising and marketing; the location of banks, malls, and industries; identifying areas at risk because of natural or technological hazards; transportation, food, and energy flows; the diffusion of regional economic development in Third World countries, optimal land-use plans and zoning in rural and urban areas; advice to political candidates running for public office; inner-city redevelopment efforts; the solutions to poverty, hunger, housing, and transportation in developing countries; the analysis and presentation of geographical and cartographical data; the delivery of health services to urban and rural areas and to the elderly; organization and reorganization of spaces (school districts, census units); and resource allocation questions. While there are many geographers working in the nonacademic world, there are many others who serve as advisers and consultants to companies and local and state governments, on national committees, and to foreign governments.

Demko (USA): I think we have a very poor idea about the use of geographers' advice and recommendations. There are a number of geographers working as consultants to government agencies and the private sector as well as many employed by these agencies. However, there must be nongeographers who read our literature, attend our meetings, and have access to geographical research. The profession desperately needs a careful and comprehensive survey of this issue.

Palm (USA): The degree to which the recommendations and suggestions of geographers are applied is to some extent a function of the position of the geographer in the contracting unit, and also a function of the degree to which the recommendations are consonant with the goals and resources of the unit. In all of the cases enumerated in response to the question about what useful knowledge and advice geographers provide to businesses and governments, for example, the recommendations of the geographers would be considered seriously and most likely adopted. Recommendations and suggestions made to a general audience, or to a professional audience in a specialized journal are exceedingly unlikely to have a major impact on the activities of business or government.

Taaffe (USA): Geographers recommendations are not widely applied in the United States. Too few decision makers have made geography a significant part of their college education, and there are simply too few professional geographers.

Turner (USA): The discipline of geography carries little weight in the applied world, with the exception of its technical elements such as cartography, remote sensing, and Geographic Information Systems (GIS). Individual geographers, however, may have enormous influence, particularly among international agencies and institutions. Gilbert White, Ian Burton, and Robert Kates, among others, have had impact on international programs dealing with man and the biosphere, hazards, and so forth. Tom Wilbanks and his cohorts at Oak Ridge have had a major impact on energy and other such applied subjects. Roger Kasperson and colleagues have influenced the application of treatment of technological hazards, especially planning for nuclear emergencies. Len Berry has set the course for resource management strategies as pursued by several United States and other national development agencies. The key point, however, is that these individuals are perceived as such, and the discipline receives very little credit indeed.

Alayev (USSR): Geographers' recommendations are taken into account very widely, but in the majority of cases, this occurs twenty-five to fifty years after these recommendations are made. That is, real-world needs demand that they be taken into consideration only when the state of the environment reaches a critical or even catastrophic level. Yet, I do not know of any environmental protection resolution regarding Lake Baikal, Lake Ladoga, or the Volga, Yasnaya

Polyana, and so forth, that would allow us to say, "Geographers warned you!"

Remembering that geography can rely on such a fundamental ally as nature and its eternal laws, geographers should more actively and courageously (and, of course, with immutable evidence) prove their point of view, their prognoses, and estimates.

Solomatin (USSR): So far, geographers' recommendations have not been taken into account by society adequately. On the other hand, such recommendations are not introduced persistently enough by geographers. So far, there have been very few illustrative examples of rational solutions based on holistic evaluations of the structures and reactions of the landscape mantle to human activities, particularly on the macroregional and global scale (e.g., river flow transfer, development of large-scale industrial-energy regions, etc.). This is owing to insufficient geographical thinking in society, and a lack of understanding of the harmonious unity of man and nature. This is caused by the fact that specialization processes still prevail in modern geography, while the synthesizing role of geography has not achieved the necessary level.

Akimenko (USSR): The question is a very important and a difficult one for geography. I would say that geographical ideas are easily perceived by society but struggle through the layers of social institutions. For instance, the ecological movement is developing all over the world. In the USSR there are youth squads for nature protection, public meetings devoted to the problem of the pollution of the environment, and so forth. Because of public pressure, the project for the transfer of northern rivers to the south was canceled, the functioning of a number of plants and factories was stopped, and other projects and decisions dealing with the use of nature are being revised. At the same time, however, the present management systems understand the complex territorial approach very poorly. In a number of cases, even knowing what should be done, they cannot always decide how to do it. This contradiction occurs because, although geosocial concerns are universal, their regulation is effected within the framework of different social institutions involved in the solution of their particular problems. Returning to the first question, it can be said that the "behavior" toward the ecosphere very often resembles a series of convulsive movements described by an old Russian proverb, "The left foot does not know what the right hand does."

Kagansky (USSR): In some spheres of activity the methods for developing schemes and projects for regional planning for environmental protection are to a considerable extent produced by geographers. But, getting planners to follow geographical recommendations is the more acute problem. Geography suggests an ever renewable picture of the laws of spatial organization of society and its environment and bases its formulations of possibilities and recommendations on this idea. But any society and or state, regulating and organizing activities in space, and implementing territorial (regional) policy, has an ideal goal for its own spatial organization. What should be done if these goals appear to be incompatible? Will society be ready to accept a version worked out by a small group of specialists if it contradicts the customary view that has a higher status? The geographer does not have the privilege that the physician has with the patient, the latter believing that the physician is better aware of the construction of the human organism than he.

Mather (USA): Recommendations and suggestions of geographers are not widely sought or applied at the present time because of the lack of adequate knowledge of what geographers do. The common perception is that geography is mostly place names, principal products, locations, and different cultures. Only when geographers themselves emphasize the useful nature of the geographic approach and demonstrate the interrelatedness of the various subsystems that make up our earth system will geographers begin to play a more prominent role in world affairs.

Morrill (USA): How widely applicable are suggestions by geographers? To be honest, not very widely applicable in the United States, except in some local areas. As we improve theory and knowledge and become more aggressive our relevance may slowly rise.

Muller (USA): It is hard to be an optimist at this point, because the recommendations of professional geographers in the United States are only rarely heeded outside of our discipline. But certain "success" stories do exist, and they should be compiled. Geographers in Washington, D.C., have much to be proud of: Harm de Blij persuading the National Geographic Society to publish a scholarly research journal; George Hoffman opening doors at the Smithsonian; George Demko and Ronald Abler working successfully to expand National Science Foundation's support of geography; Julian Wolpert, Brian Berry, and Robert Kates gaining visibility for geography at the National Academy of Sciences; George Demko overseeing the expansion and rising influence of the U.S. Office of the Geographer at the State Department; Richard Morrill serving the Washington State Supreme Court as a Master for redistricting that state. The academic careers of Frank Horton and other geographer-administrators are important. There are a number of individual achievements worth noting, but, in general, the situation could certainly improve.

What is required to improve and expand the application of geographical knowledge and approaches to human problems?

In order to improve and expand the application of geographical analysis to human problems, Soviet and American geographers again offer similar suggestions. There is a clear consensus that geographers must publicize their skills and research results more aggressively and widely, especially to government agencies. The need for more and better training for advanced geography students was similarly recognized. All United States geographers strongly argue for more attention to, and training in, applied geography for students that would better prepare them for practical roles in addressing and resolving societal problems. Also, suggestions regarding reorganization of the profession to concentrate research efforts are made. A number of Soviet geographers recommended the creation of a new and separate state committee or analogous institution for geography that would represent the profession in the large array of national projects.

Khorev (USSR): It is important that geographers' recommendations be reckoned with to a greater extent than they now are. In our era, geographers often honestly and insistently, although, perhaps too timidly, warn others that the level of our knowledge about natural-geographical processes does not permit bold experimenting with the environment or with large-scale interference in natural systems. Modern technology is powerful, and it is very tempting to alter, for example, the flow of rivers or alter the diversity in local relief of an area even though we comprehend poorly the consequences of such actions. And disastrous impacts to entire systems may result. As long as it is possible to rationally "blend" humanity with the existing natural environment, it is too early to tackle large-scale projects to "remake" nature. It is a question of principle, if you will.

In order to create a reliable monitoring system for the environment and to learn to "control" individual processes, it is necessary, first of all, to create a reliable organizational base. Therefore, we need a state geographical service for the country, uniting, for example, such departments as the State Committee for Hydrometeorology and

Natural Environment Monitoring, and the Main Department of Geodesy and Cartography. It is worth noting that hydrology and meteorology, geodesy and cartography are, in general, components of geographical science, and they are represented in geography departments of higher educational institutions of the country at the level of subfaculties. At any rate, a unique governmental body dealing with questions of environmental protection and monitoring is imperative. There are already such committees (Goskomprirodi, the State Committee for Ecology) in individual republics of the country. It is all necessary to develop and form a new scientific discipline, ecological geography, and create training and research units in the field.

New life should be pumped into the cause of saving nature, which was started by geographers some time ago. Comprehensive surface monitoring of environmental variables is feasible mainly in large cities and particularly vulnerable localities. It seems, however, that we need a cadastral survey of the environment, a kind of "red book" for the future. The scientific community already feels the need for such a cadastre. For example, the Pacific Ocean Institute of Geography of the Far East has developed and published Methodological Recommendations for Compiling Cadastres of Protected Natural Territories of the Far East (Vladivostok 1986).

Geography and geographers should tackle this problem as well as take responsibility for a kind of public service for protecting nature and rationally using the environment in general. It is also necessary to develop more active research at the interfaces of social, natural, and technical sciences where discovery of new knowledge is possible. Geography is in a most advantageous position at this time, and its contribution can be increased, particularly, regarding research into problems of environmental use, spatial organization of society and the environment, population, settlement systems, urban planning, and regional policy for science and technology. Such special geographical fields as urban geography and ameliorative, medical, and recreational geographies are well advanced using experience from the interface of sciences. It is time to apply geographic technology as a priority research effort.

Grishankov (USSR): For centuries people gradually learned to understand and exploit the peculiarities of natural conditions in their economies. At first, understanding of natural laws and economic activities were in unison. The peasant was a tiller, a meteorologist, and a botanist. Economic activity was, on the one hand, an activity meeting the requirements of people while, on the other hand, it was a scientific experiment allowing certain natural laws to be discerned. In the late 1800s, early twentieth century, economic and scientific cognitive activities separated, and reuniting them became increasingly more complicated.

In the USSR at present, questions of applying the results of geographical research to social production and problems have been, to a degree, resolved. The relationship of geographical sciences to production is realized in the following system: research institution --- state bodies --- economic enterprises; or: scientific institutions --- economic enterprises. Heads of enterprises (plants, collective farms, state farms), engineers, agronomists, and so forth, are, in most cases, the conductors of new ideas. Under such conditions, however, specific geographical recommendations have taken into account the degree to which these specialists understand the essence of geographical laws. Therefore, to improve geographical competence, some geographical training such as short courses in meteorology, hydrology, and geomorphology, have been introduced in a number of higher educational institutions. This method, however, is inefficient in terms of transferring the advances of analytical geography to economic production, and it does not solve the problem of using geographical knowledge regarding environmental complexes. Examples of negative consequences resulting from incorrect knowledge about regional problems are well known --- inappropriate introduction of grassland crop rotations, the damming of Kara-Bogas-Gol (Caspian Sea), the problem of Lake Baikal, and a number of others.

With the current specialization in science, it is clear that one specialist cannot be a master of the entire scientific system or even the bulk of it. Thus, chemists solve chemical problems, biologists solve biological problems, and so on. A strange attitude, however, has developed in a number of sciences, which implies that scientists can solve their own problems. Pedagogical science and physical geography are among them. The simplified conception of physical geography as a science is characteristic of many parts of the economic system. As a result, neither economic enterprises, nor state agencies have specialists capable of incorporating data from the geographical sciences, particularly the landscape sciences.

The establishment of wider and closer ties between geography and social-economic production requires a more active dissemination of geographical knowledge among the population.It would be advisable to publish a popular magazine on geography and the economy, and to set up a state geographical service that would become the unifying link between geography, state agencies, and economic enterprises.

Kagansky (USSR): The geographical community itself should make it clear how its knowledge should be disseminated so as to be noticed and used. In order for society to be ready to use geographical knowledge, it must have a need to apply the concept of spatial organization and the desire to understand the world's real diversity. A socially conscious public should value its space, the cultural landscape as a whole, not just its individual preserves or monuments. The public, having noticed the existence of geography, should then recognize the sphere of its competence.

Brunn (USA): Geographers need to recognize that we have the appropriate concepts and techniques to study global problems, that it is important that we address major human problems, that we must actively promote efforts (training, retraining, funding, publicity, awards) that call for such increased understanding, and that we identify (by consensus) priorities for our research for the next decade. All four are important and realistic solutions. There is no question in my mind but that geographers are equipped (perhaps uniquely because of our holistic perspective on the world) to undertake the study of major human problems, whether that be nuclear winter, climate change, environmental modification, natural hazards, resource depletion, hunger, cultural and social conflict, or the delivery of vital services to the poor, rural, and elderly. That awareness both of problem identification and resolution needs to be conveyed to our students, in our conferences, in our journals, and in research funding. To accomplish such goals will likely call for specific short, medium, and long-term efforts by the discipline's practitioners (in universities and the public and private sectors). We must train our young geographers, with this renewed focus in mind, to offer our services and expertise to those agencies, corporations, governments, and other institutions interested in resolving human problems; and to make ourselves available to cooperate on national and international teams investigating major and environmental problems.

de Souza (USA): Our responsibility as scholars and teachers is to try to understand the world as it exists for all humankind and to explore the causes of that reality. The world presents challenges of immense importance and complexity. If we are to improve and expand the role and responsibility of the geographer in studying human problems, we should explore our own world view, a theme in the history of geography and geographical ideas. American geographers need to examine their assumptions and theories. For example, in their work on development problems of underdeveloped countries they often use models that commonly work counter to what is claimed for them.

Cohen (USA): The biggest problem is to attract high caliber graduate students, well versed in a cognate discipline, who will compete at an intellectual level with the younger generation entering other disciplines.

Taaffe (USA): The combination of small size and enormous diversity poses a problem for geography in that we have difficulty achieving sufficient scholarly penetration into any given subtopic or problem area. There are seldom enough competent investigators available per subtopic to achieve a critical mass. One answer is specialization on the part of the graduate research departments. Such departments should not try to be all things to all people but should focus on the kind of work in which they feel they can meet the highest standards of the scholarly community. Centers of excellence would represent an extension of this idea. The organizational and financial difficulties of developing such centers, however, may preclude or delay the implementation of this alternative. In the meantime, a steady narrowing of focus and concomitant elevation of standards on the part of graduate departments is quite possible and need not await the development of a master plan.

A second need in the expanded application of geographic work to human problems is to establish a closer link between theoretical work, on the one hand, and applied as well as regional work, on the other. Here, however, I would reverse the usual exhortation, which seems to be mainly addressed to the theorist to make his work more applied. Instead I would argue that it is the applied geographer and the regional geographer who should extend the greatest effort to bring emerging concepts and techniques in spatial or environmental analysis to bear on a particular problem or region. The conceptual apparatus is more readily available in the literature to the applied geographer than the institutional and historical complexities of a particular problem or area are to a geographer whose greatest skill is in the development of new theoretical or analytical approaches to such problems.

Tobler (USA): One way to improve and expand application of geographic knowledge may be with more practical knowledge concerning settlement sizes. Most important, however, is more and better theory.

Palm (USA): Applied geography will have a greater impact to the extent that contracting agencies specifically ask for input from geographers. The more consulting or contract work done by university geographers, or the more geographers are directly employed by government or private industry, the more their work will directly affect the activities of government and industry. A greater awareness of the potential contributions of geographical analysis through more attention in the popular media, as well as a more effective undergraduate training in geography, will enhance the relationship between academic geography and the private and public sectors.

Willmott (USA): If geographers (as a group) had a better grounding in science, math and computer methods, they would be much more credible. The solution, consequently, lies in upgrading the training of geographers. To communicate effectively with the scientists, engineers, economists, and so forth, who "run the show," we must become well versed in biology, chemistry, physics, math, computer methods, statistics, and so on. Without such knowledge, we will be increasingly relegated to "ivory-tower" discussions and trivial problems.

Morrill (USA): Expanded application of geographical expertise to societal problems is a function of cumulative effort --- better education, including explicit training in practical models and techniques; more publicity on good geographic applications; better organization in its delivery (in the United States, geographic education and research is too individual and dispersed); we need some centers or institutes of more concentrated effort.

Mather (USA): To improve and expand the application of geographical knowledge to human problems requires a greater awareness not only among the educated public and government officials but also among geographers themselves of the unique nature and contribution of a geographical approach. Many geographers do not fully understand their own field and so persist in the stereotypical view of a limited (nonintegrative) approach to problems. Thus, both better education of the public via our high schools and colleges as well as an increased awareness among geographers themselves of the real contribution that their discipline can make will be necessary to expand the application of geographical knowledge.

Demko (USA): In order to expand and improve the application of geographical knowledge to human problems more geographers must address such problems in their research and in the training of geographers. Too many of our scholars are focused on unimportant or trivial issues that may be personally interesting and rewarding but are irrelevant to society's needs. In addition, geographers must communicate with individuals in organizations that deal with pressing, societal problems, and work with government agencies, private institutions, and so forth, to determine the issues, acquire the data, and understand the complexity and importance of such problems.

Lavrov (USSR): What should be done to apply geography to societal problems and thereby increase its social status? In my opinion, primarily "concreteness" of research is needed as well as "addressing" the right group or problem. Research should be oriented to those areas where there are real management institutions capable of implementing the recommendations of science. Secondly, the complex character of research is important and should be aimed at the entire chain of physical, economic, and social consequences of human activity rather than some narrow aspect of life (in this, lies the power of geography). Thirdly, the scale of investigation should be greatly increased. Research should not avoid acute problems of regional or interregional projects ("transfer of rivers," the problems of the Caspian Sea, Aral Sea, Baikal, etc.).

All this relates to the essence of research but, there is another problem, larger and more unusual. It is necessary to advocate geography, showing, via the widest use of the mass media, its potential, "the growing importance of the geomethod " (according to the words of the late outstanding philosopher, academician B. M. Kedrov).

Some proposals regarding this question include the following:

1. Increase the level of teaching geographical knowledge (particularly in the field of nature conservation) throughout the entire educational system.
2. Include geography (geographers) in the structure of an organization dealing with project inspections as well as in into the decision-making organizations (for instance, a regional department with a "head geographer"). This would make it possible to centralize all measures to protect the environment and create a primary client for developing a territorial complex scheme for nature conservation, regional planning, and the development and deployment of productive forces.
3. Geographers should tackle more courageously state plans and orders where geographical knowledge is a determining factor for success.

O. A. Evteyev (USSR): The following recommendations would help. The development of geographical research directly connected with territorial organization, management, planning of economic and social activities (within the framework of the country as a whole, its regions, and individual localities), including the geographical inspection and review of projects. We should also improve geoinformation flow from science to application and create automated geography-oriented data banks, and series of maps and atlases at global, national, regional, and local levels.

Finally, we should develop methods of geographical investigation including mathematical, remote sensing, cartographic methods, in order to provide geographers with adequate technology.

Bokov (USSR): In order to expand such applications we must increase the level of research. One of the conditions for achieving this is better training at universities in subjects connected with the rational use of natural resources (engineering geography, geographical reviews, optimization methods, etc.). In addition, leading scientific geographical groups should, via convincing examples, demonstrate the potential for geographers to solve societal problems. It would help to convince planning bodies to introduce the job of geographer to the staffs of organizations and enterprises.

CHAPTER FOUR

Geographical Approaches to Contemporary Global Problems

How can geographers contribute to the resolution of environmental problems?

The range of suggestions and comments by Soviet and American geographers on the question of how the profession can contribute to the resolution of environmental problems is great and varied. All respondents note the need to have geographic work and research results more visible and accessible to decision makers in both societies. The special contribution that is possible using geographic approaches --- regionalization, spatial analyses, and so forth, and the application of geographic methods, especially cartographic and computer data bases, is described by respondents in both countries. American geographers also recommend that more emphases be placed on environmental issues in graduate training programs and that geographers become more active in international formulas dealing with environmental issues.

Karpov (USSR): The eternal problem of the interaction of society and nature and its regional patterns and general laws has always been the focus of attention of geography. This problem has an unusual qualitative aspect, has acquired new content, and has become more complicated, touching on and directly affecting both the natural state of the world and various aspects of the material, social, and spiritual life of contemporary society. But, the most important aspect of the present state of the environmental problem is its growing global significance. The pollution of the environment is a problem of global scale, and the demographic situation also has global ramifications and is attracting international attention as are famine and malnutrition in many areas and the backwardness of developing countries. Through the efforts of all society such contemporary problems as the development of outer space, world ocean resources, and the provision of natural resources for economic, recreational, esthetic, and other purposes for all people can be solved.

What are the tasks of geographers in the solution of such problems? In addition to the solution of routine and specific questions in which geography is traditionally involved, we should find solutions for a part or many of these contemporary global problems. They can roughly be defined in the following manner:

1. Global problems are known to be interdisciplinary problems requiring study by different sciences. Geography, being one of the most synthetic sciences, one that unites both natural and social science, could be such a foundation and body of experience upon which a methodology for researching global problems could be developed.
2. The creation of an information base for global problems is to a degree already based on research by geographical scientists. The compilation of cadastral surveys, maps, field observations, and related data within the framework of geographical research can be extended and modified to consider information required for solving global problems.

Most global problems, of course, have regional and national impacts. The experience analyzing interconnections between national and regional scales and general and local regularities by geographers may also be used to study global problems. Essentially, the creation of a new branch of geographical knowledge is making it possible to establish distinct interconnections between regional manifestations of global problems and their common characteristics, and to reveal the genetic, territorial, and other features of these manifestations is the geography of global problems.

These proposals do not, of course, exhaust the possibilities geographers have to solve vital contemporary problems. They might, however, become the basis for a systematic approach and for the concentration of efforts by geographers who have been working for many years in this direction.

Treshnikov and Lavrov (USSR): We believe that any global problem cannot be solved without previous "regionalization," without detailed consideration of national and regional scales, and determining specific solutions for particular places and conditions (natural, economic, and social).

Let us recall that the first global models (Meadows) were justifiably criticized for their exclusive globality and the absence of "regionalization." The models of M. Mesarovitch and E. Pestel showed the world as a "collection of regional models" rather than a realistic analysis of all the diverse problems and tendencies of development of different systems and regions.

As "globalistics" developed, the significance of the regional approach and the "geographization" of the models increased, which was natural. For example, the American model, Global 2000 (1980), was divided into sectors --- population, social production, climate, technology, food, agriculture, fishing, forests, and so forth --- and concluded with a serious ecological generalization. The justifiable conclusion was, for instance, that there is no "water economy of the world as a whole," for water resources are, in essence, regional systems.

Such an approach seems to be valid to apply to the ecological situation in any individual country as well, particularly, large countries such as the USSR or United States.

What is the role of geography in solving ecological problems? Ecological problems are interdisciplinary and, to solve them, it is necessary to integrate natural, social, and engineering sciences in one complex. In this complex (some refer to it as social ecology and others as human ecology) geography should study regional ecological specificity and regional linkages in the system, nature-production-society. In the West, the name geoecology for this portion of the interdisciplinary complex has been adopted. It is also being accepted in some socialist countries such as the GDR where the primary institute within the Academy of Sciences is called the Institute of Geography and Geoecology. In the Soviet Union, territorial complex schemes for nature conservation are compiled at all regional levels ranging from the oblast, a territorial administrative division within a republic, to larger divisions. It is interesting that the methodological basis for this scheme in Estonia was provided by geographers at Tartu University. It is important to note that a set of original maps (maps of technogenic pressure, conflict situations, and conservation measures) represents the most adequate and understandable reflection of the ecological situation for the broadest circle of "users."

The West recognizes that in the USSR everything is decided by ruling bodies. Meanwhile, the most important decisions in the field of environmental protection were made despite the opinion of many ministries and departments. For example, the most important decision to stop the work on the transfer of river flows (1986) was made because of the opinion of the public and scientists, geographers in particular.

We believe that the problems of ecological education can be solved only by geography departments at leading universities of the country. It is quite evident that ecologies will be needed in the future that will be able to integrate data from other sciences and other specialists to provide competent ecological forecasts and the development of a set of measures for large regions or urban agglomerations. This is part of that very spatial organization which, in every part of the world, is considered the task of modern geography.

Although there have been many unexpected and unforeseeable events of late, ecological consciousness has not yet appeared. Perhaps, had it not been for the implementation on the 1930s of the policy of redistributing productive forces in the USSR to the East, as well as the policy of constraining the growth of large cities (with geographers participating), the ecological situation in the European part of the USSR would be far more complicated today.

We believe that geography uniting both natural and social fields is becoming the main field for ecological research.

Kondratiev (USSR): The strategy of accelerating socioeconomic development of the country, which became the basis of all our work after the XXVII Party Congress, requires serious attention to interdisciplinary problems of the natural environment closely connected with the progress of the national economy and the social sphere. The scientific and technological revolution has pointed up the interconnectedness of natural and social systems on a global scale. The most important aspects are:

1. the concept of certain mutual annihilation proclaimed by western strategists as the main means of equilibrium between the opposing social systems;
2. the acute problem of foreign debt in developing countries;
3. the crisis level of some impacts on the biosphere and the environment, related in particular to the barbarous exploitation of natural resources (for instance, the destruction of tropical forests is already producing changes in the global carbon circulation, which affects climate), uncontrollable pollution of the atmosphere, and crossborder transfer of pollutants due to atmospheric circulation.

An important new feature of our era is the global ecological threat to the world community caused by imperialism. Nuclear war as the source of a global ecological catastrophe, as a threat to the immortality of humanity, is the main component of this threat.

The interdisciplinary nature of the problems of the biosphere and the environment requires closer cooperation by specialists in the natural and social sciences. In this connection, there is an acute need to analyze the connections within the noosphere put forward by V. I. Vernadsky, as well as prospects of humanity becoming autotrophic. The experience of the recent past shows that none of the major problems of natural science can be solved without a systematic, interdisciplinary approach. Speculative concepts that bring misinformation are the only alternative in this case. We can recall examples of apocalyptic forecasts of climatic change by the middle of the next century that were void of any serious scientific substantiation, as well as the similarly gross errors of other ecological prognoses (e.g., the Caspian Sea level, the decision to dam Kara-Gogaz-Gol, and the predicted consequences of the transfer of part of river flows). This confirms the urgent need for a serious discussion of the concepts regarding the quality of the natural environment and the need to make them more concrete.

The need to reconstruct our consciousness demonstrated by the documents of the XXVII Party Congress (without which radical changes are impossible) plays an important role in investigations of natural environmental development. The efficiency of the human factor enters as a decisive condition for the solution of problems of such extreme complexity and vital significance.

Global environmental problems can be solved only on the basis of efficient international cooperation. One of the most promising possible areas for cooperation is the implementation of the International Geospheric-Biospheric Programme (IGBP), approved by the XXI General Assembly of the International Council of Scientific Unions in September 1986. The main objective of the IGBP is to describe and understand the interacting physical, chemical, and biological processes that regulate all earth systems (especially the unique natural environment which provides for life) and the changes taking place in the system and the nature of man's impact on such changes. There are the five main research directions: (1) the reconstruction of the changes in the past from data for different paleoindicators; (2) the study of different changes taking place in the natural environment at present; (3) analysis of global changes in the biosphere; (4) the study of biogeochemical circulation; and (5) the study of the global circulation of water.

It seems that, in terms of the contemporary concepts of processes in the global geosphere-biosphere, ecological investigation should proceed while taking into account: (1) socio-natural-scientific approaches to the problem; (2) verification of such strategies as controlling geospheric-biospheric processes that provide the harmonious interaction of human economic activities with nature, and preserving (recovering, if necessary) the equilibrium of the circulation of matter in nature.

Accumulated observed data testify that the rate of change of the mass of main chemical elements available to be used by life forms to the change of the mass due to synthesis (or decay) of organic matter is of an order of ten years. Without equality between synthesis and decay flows, the chemical composition of the natural environment will change, and all life resources will be exhausted in about ten years. The existence of all living organisms, including man, is stable because their interaction provides for the closed biochemical circulation of substance in chemically invariable environment. The period for species formation is not less than 10^5 years. Cycles of synthesis and decay of organic matter of stable species during this period should coincide with a relative accuracy of about 10^4, a conclusion supported by paleodata. Such high accuracy can be provided only by certain communities of interacting species. Research on the biology of organisms forming such communities is urgently needed and a fundamental problem of biophysics and ecology.

It is expedient to try to determine how problems of a global and regional scale are interconnected. Global problems should be dealt with by evaluating not only the present state of things but the prognoses of the evolution of the geosphere-biosphere during the next decades and centuries. The main content of regional problems is in the analysis of observed impacts of economic activities on nature (ameliorative measures, hydrotechnical constructions, and pollution of the natural environment), which takes into account their priority and develops and uses simulation modeling to estimate effects on the environment; verification is required to determine ecologically safe technologies. In any case, the philosophical, political, social, and economic aspects of problems are important components.

There are, among the main objectives of ecological research, fundamental investigations with the following tasks: (1) understanding the nature of the biosphere as a system providing life support over long spans of time; (2) understanding the role of the interaction of animate and inanimate matter in nature; and (3) analyzing conditions for the survival of the biosphere and its resistance to external effects.

The key importance of geography for solving many of the above problems is quite obvious.

S. A. Evteyev (USSR): The contribution of geographers to the solution of global ecological problems can be very great. Environmental problems arise because the peculiarities of the environment and its reaction to anthropogenic pressure are not adequately considered in the process of socioeconomic development. Hence, the potential solution to ecological problems, including global ones, is through the rational use of nature, which should consider the viability of natural ecosystems and their ability to reproduce natural resources. As to the nonreproducible resources, the problem is to use them efficiently and so as to minimize, in any specific case, the extraction of the nonreproducible resource and the volume of waste that becomes pollution in the biosphere.

In order to work out a long-term rational ecological strategy that is not economically contradictory, it is necessary to know the history of ecosystem development, its present functioning under anthropogenic stress, and to be able to make predictions of natural changes in the process of socioeconomic development. Undoubtedly, we need a complex body of information and knowledge about natural and socioeconomic processes. Which of the contemporary sciences is capable of contributing greatly to the solution of all these problems? Some think that this should be done by a new, complex, integrating science that allegedly can deal with the study of the entire gamut of interactions between society and the environment; global ecology, general ecology, megaecology, nooecology, and social ecology claim the role of such a science. This is erroneous from our point of view. Current ecological problems encompass a wide spectrum "from geology to ideology." Their solution cannot be packed into the framework of one science, no matter how widely its limits could be extended. We can and must consider what contribution a particular science can make, in terms of its focus and research methods, to the solution of environmental problems.

The role of geography is extremely important. At least two factors explain the priority of geography. First, it is a science which organically combines the study of natural and socioeconomic phenomena. A second strong point has always been that our study of land is understood in its broadest meaning, as ranging from complex studies of the territorial-industrial (agrarian) complex to the region and globe.

Environmental problems are not static. They change over space and time and are always linked to certain territories with a new, often unique, set of natural and socioeconomic characteristics, which makes the role of geography in their solution important.

We believe that at present geography is far from having exhausted its potential for solving environmental problems, because geographers participation in solving important economic problems connected with the rational use of nature is not enough. In particular, in our country, the participation of geographers in working out principles and methods for compiling territorial complex schemes of nature conservation, important pre-plan documents considering the close interaction of socioeconomic development and environmental protection, and the preservation of the viability and productivity of the environment is not sufficient. Other examples could be given.

All the above makes it possible to conclude that modern geography can make a great contribution to the solution of environmental problems, including global ones, but it should resolutely turn attention to the problems of rational, ecologically verified use of nature.

Trofimov and Solodukho (USSR): Different methods --- technical, political, legal, and sociomoral --- are applied in solving global ecological problems. Methods of global modeling and forecasting including geosituational analysis will play an important role in their solution.

Geosituations (corresponding to ecosystems) are local inhomogeneities, such as concrete states of the global ecosystem (man-environment) that appear and disintegrate, exist, function and develop, and decelerate or accelerate the form, function, and development of other geosituations.

For the control of ecosystems to be efficient, they should be based on modeling geosituations (ecosituations), as finer and more flexible mechanisms of change than on modeling geosystems. The geosituational approach makes it possible to determine the beginning phases of forthcoming changes in the ecosystem and consequently makes the control of its development more effective.

It is easier to correct an undesirable ecosituation, if necessary, than to influence an already altered ecological system that has undergone irreversible qualitative changes. The above discussion refers to global ecosystems, the control of which should be carried out primarily at the lowest levels of the geofield, regional and local. It is at these levels that geosituations leading to fundamental changes at the global scale germinate and accumulate as local inhomogeneities of a geosituation. Control of the global ecosystem should start with these inhomogeneities.

It is easier to use natural resources rationally than to find new sources of ores, energy, soils, and fresh water. It is easier to protect the earth, atmosphere, and oceans than to purify them after irreversible changes due to anthropogenic "contamination." The geosituational (ecosituational) approach aims at revealing what is easier to do in the geosystem (ecosystem) before irreversible system changes have taken place.

Vikulov (USSR): It is impossible to overestimate the contribution of geographers to the solution of ecological problems at any level, from global to local-specific. Geographers must, in most cases, be first (and actually are), both in raising ecological problems and in solving them. This is explained by a number of circumstances.

First, geographers as representatives of the natural sciences possess well-developed ecological thinking that is sharpened as a result of the accumulation of experience in the process of working in the field.

Second, because of professional peculiarities, geographers constantly observe, and note actual changes taking place in nature, including those caused by anthropogenic factors. Geography is constantly accumulating information on changes of natural complexes of any rank, as regular geographical investigations cover all the land and water surface of the earth from pole to pole, as well as the near cosmos. A peculiar geographical monitoring (to be more exact, complex monitoring) results.

A third circumstance is revealed in the notion of "rational use of nature" which is seen as the only way to solve ecological problems. (We shall not consider here the social component of the term and its transformation in time in the process of the developing productive forces.)

Rational environmental use is based on a tripartite basis --- ecology, technology, economy - that is based on observation of ecological requirements, technological possibilities, and economic expediency. Geography has the first priority in solving two questions of this triad --- ecology and economy. Geographers must estimate the adequacy of a particular enterprise or a particular project in terms of both the ecological aspect and economic aspect. The opinion of economic geographers is decisive in questions of locating a particular enterprise.

Geographers should clearly be aware of their primary role in solving ecological problems, and each of us should carry it out as well as possible in his work.

Mather (USA): By emphasizing the interrelatedness of all of the subsystems (atmosphere, ocean, biosphere, and human-economic) that ultimately make up the ecosystems of the earth, geographers can contribute to the resolution of the many existing environmental problems. Environmental problems should not be evaluated in isolation, without full consideration for the interconnections and feedbacks among related subsystems.

Morrill (USA): With regard to environmental problems and following from earlier arguments, geographical analysis has the advantage of recognizing the spatial interdependence of physical and human processes that affect the environment. Among other contributions, we can help define the appropriate scale at which to study and resolve problems and help to distinguish the relevant from extraneous factors.

Willmott (USA): Basic research is needed to uncover the important underlying geographic processes. Right now geographers should be active in several large-scale research areas, for example, global climate modeling and global economic modeling. Relationships between artificial intelligence (AI) and Geographic Information Systems (GIS) will, no doubt, also become very important in the not-too-distant future.

Scientific geographers should serve on national and international committees charged with outlining directions for funded research and with making recommendations to agencies responsible for environmental protection and conservation.

Palm (USA): Geographers can contribute to the resolution of environmental problems both as scholars and as world citizens. In the realm of scholarship, geographers can help to provide information about the impacts of human activity on the environment and a better understanding of environmental processes that are beyond the control of human society. As citizens, geographers can press for greater attention to the system-wide impacts of human activities --- particularly of the so-called technological hazards that threaten to have major impacts on the global ecosystem. Geographers, along with other physical and social scientists, should participate in major national and international initiatives to understand, monitor, and intervene in the global ecosystem. Several projects of this nature are currently underway, some under the sponsorship of such international bodies as IASSA (Institute for Applied Systems Analysis) or the United Nations University.

Taaffe (USA): In addition to the obvious recommendation that more people study geography, ecological research itself should be focused more on developing cumulative findings and on placing more emphasis on the relations between individual research studies. In ecological or man-land research there seems to be few general conceptual themes available to guide research and ways to approach long-standing environmental problems. Ultimately, it is to be hoped that ecological research by geographers and others will provide a better understanding of two of society's major problems: the impact on the environment of changing technology and the nature of the long and complex causal chains that characterize environmental phenomena.

de Souza (USA): The concept of environment is central to geography, and human-environmental relationships are among the greatest long-term crises facing humankind. Unfortunately, geographers, who were embarrassed by their espousal of environmental determinism in the early years of the twentieth century, largely abandoned research on the environment until the energy and resource crises of the 1970s. As a result, geography is now one of several disciplines engaged in environmental research.

If geographers are to contribute strongly to the task of resolving environmental problems, they must focus more than they have on using geographical approaches. In addition, geographers must pay more attention to the fact that environmental problems can be viewed from different vantage points and, consequently, efforts to resolve them can be approached from a variety of perspectives. Finally, geographers should approach the study of environmental problems through a sequence of scales ranging from the local or regional level to the global level. Local environmental problems are often a consequence of decisions made by officials at national and international levels.

Turner (USA): It might be argued that the nature-society component of the discipline of geography has developed a good understanding of environmentally related issues. In general, however, global issues such as these are best addressed by outstanding synthesizers, and the latter can develop in all disciplines.

Geographers can contribute, as they have, by examining the subjects in question in their particular ways and making the results of their work known to a wider audience.

Brunn (USA): Geographers can contribute to the resolution of such problems by examining more closely than we have both the human-environment interfaces, designing innovative and appropriate methodologies to study those interfaces and interrelationships, focusing more closely on spatial public policy dimensions, utilizing innovative techniques and technologies (computer cartography, geographic information systems, and artificial intelligence), and availing ourselves and our colleagues of opportunities to work with others to resolve problems. We need to be assertive and aggressive in these efforts. Also, I believe it is important to develop a regional awareness of environmental problems and present a regional perspective to agencies and governments. Much more work needs to be done in the international arena and in a cross-cultural context, not only in behaviors, perceptions, and planning, but in public policies. Also geographers can contribute, perhaps better than others, to understanding the interfaces between the natural and technological hazards. The scale, location, and cartographic dimensions must be recognized and stressed as important ingredients. Finally, I believe geographers may be better predictors and forecasters of environmental problems than scholars in economics, biology, or the earth sciences.

Demko (USA): Geographers are already contributing to the resolution of environmental problems to a large degree. Geographers such as Gilbert White and Robert Kates in the United States and M. S. Kotlyakov in the USSR are internationally known and called upon for their expertise and knowledge about environmental issues. Geographers at Clark University have organized large, successful, international conferences on environmental problems and are publicizing the results widely to other scholars, disciplines, and the public.

Much more is needed, however. Geographers must communicate and cooperate more with each other and build critical masses of researchers to address major issues. Too often individual geographers work in isolation from each other because of funding demands by universities or other research institutions that are more concerned with their own economic health than with the issues to be studied.

Dubinsky and Riman (USSR): The contribution of geographers to the solution of global ecological problems should be accomplished by introducing optimal programs for working with nature in every situation that must consider regional features. If such a program is implemented in each case, the solution of global ecological problems will become much more feasible. The importance of constant contacts with geographers from different countries at different levels is important also, with a view to solving problems beyond any regional framework. In this case as well, placing everything in order within one's own set of natural-territorial complexes and keeping negative ecological consequences from spreading to neighboring territories would be a considerable contribution of each side to the solution of the ecological problem.

Rodoman (USSR): If preventing war, world resource-economic crises, and the impoverishment of many developing countries can be considered global problems, then geography is closely related to the two latter issues and provides much concrete material for their understanding. What is needed is concrete knowledge of lands, rather than prognostic thinking. The classical geographer's erudition is capable of evaluating, reminding, and warning. The greater the hope for an agreement on arms limitation, the more threatening seems another hazard, an ecological crisis. And, ecology is now a field for all geographical sciences. It is said, however, that no civilization perished as a result of foreseeable events. This may console one and raise hopes that certain global problems are transient. On the other hand, it is frightening. Some decades ago it was impossible to foresee the AIDS pandemic or the outbreak, on an international scale, of Islamic fundamentalism. Is the source of future storms similarly unexpected? Does this mean that the sky over our heads will never be cloudless? Global problems require that people be broadly educated as never before, that they be capable of original thinking and of working easily with diverse material "ranging from geography to ideology." What other field of the humanities is more trained for this? I would place geography second after philosophy.

What can geographers suggest for the preservation and more rational use of the natural resources of the earth?

Soviet and American geographers, addressing the question of more rational use of natural resources on the earth, agree that their research and views can be very useful and productive. They suggest that geography's man-environment research traditions, sensitivity to the interdependence of regional systems at all scales, and geographers' spatial data systems are invaluable inputs for resource study and planning at international levels. Soviet colleagues describe a number of specific research areas relating to energy and other resources for geographers.

Brunn (USA) : The concerns of geographers for the preservation of the environment and a more rational use of natural resources can be expressed in three areas. First is the long-standing human-land (or sometimes called man-land) tradition. In this paradigm, which is subscribed to by most cultural geographers, there are concerns about human use and misuse of the earth, human modification of the earth's natural environment (climates, land form surfaces, vegetation, and soils), and the conservation of planetary, natural, biological, and human resources. The schools of cultural ecology, cultural adaptation, and landscape modification are expressly concerned with furthering our understanding of the human-land ties in traditional and contemporary societies in both developed and developing countries.

The second area of study that is associated with environmental preservation and natural resource issues is that of humanistic geographers. To these geographers the concerns are the values that societies attach to land, the experiences that individuals and cultures have with land, and the meanings of land, landscape, and environment to a culture's religion, patterns of economic livelihood, and prevailing ethics as reflected in laws and regulations protecting the earth's resources. Geographers interested in these questions may find the human expressions of nature, environment, and human relations with the environment addressed in art, music, and theater, as well as in a culture's religion, philosophy, and leisure pursuits. Related work by anthropologists, historians, philosophers, and literary critics will be sought by humanistic geographers. The third area of study includes those geographers interested in social geography, broadly defined, and in particular with public policy and social welfare questions. Included under this rubric are concerns about geographers responding to contemporary social issues and current problems. Environmental conservation, the preservation of the planet's species, prudent use of natural resources, and the future of biosphere are major concerns of many social and political geographers. Their methodologies may be descriptive, analytical, or prescriptive (forecasting) and may include examining the above problem at a local or global scale. These studies draw support from related work by resource economists, biologists, political scientists, environmental psychologists, and social-environmental engineers.

Sokolov (USSR) : It may not be an exaggeration to say that water problems occupy a central place among resource issues. Population growth, development of economies and cultures, the process of continuous increased demand for water in industry, and communal and agricultural economic growth are going on all over the world. Every twenty years the volume of water used almost doubles. Simultaneously, the rate of discharge of water into rivers and reservoirs that are traditionally considered as the last stage of the sewerage system also doubles. This is, allegedly, where self-purification of water occurs. This result, in many countries, is rapid exhaustion of water resources, quantitatively and, particularly, qualitatively, and the violation of the ecological balance of nature established over centuries. The water problem goes beyond a national framework and takes on an increasingly global character.

In 1965, to stimulate the study of water resources and the water balance of the earth, UNESCO established an unprecedented, large-scale, international hydrological program that has been operating for more than twenty years. More than one hundred countries from all continents are taking part in its activities.

Soviet scientists from four major research institutions --- the State Hydrological Institute, the Main Geophysical Observatory, the Research Institute for the Arctic and the Antarctic, and the Institute of Water Problems --- have participated, a national contribution by geographers to the fundamental study of the circulation of water on the earth and its links --- the atmosphere, the hydrosphere, and the lithosphere. We have summarized the latest data on the stores and resources of all kinds of water, their geographical distribution over the continents, river basins, and countries.

As a result of the study, it was shown that, based on the growth of population and productive forces, and even with strict water conservation measures, water use will reach 6,000 km^3 per year by the year 2000. Mankind is faced with the task of rationally using and preserving the water riches of our planet.

Research materials were published in 1974 as the monograph, *World Water Balance and Water Resources of the Earth* and the *Atlas of the World's Water Balance*, containing sixty-five maps on all elements of the water balance of all continents. They were recognized all over the world, and in 1981 these works by Soviet geographers were awarded a state prize.

Shcherban (USSR) : Proceeding from the inevitability of the exhaustion of mineral energy resources, geographers persistently study the spatio-temporal distribution of such ecologically limitless natural energy resources as solar radiation, wind, ocean waves, and thermal waters. It is true that thermal waters are not always ecologically pure, since at high temperatures there is the possibility of many active (aggressive) chemicals being dissolved in them and their presence is capable of causing undesirable effects on the soil-vegetation cover, animals, and man. However, the presence of iodine and boron compounds in the thermal waters of Iceland allows them to be intensively used for pharmaceutical purposes.

The energy of ocean waves has not been adequately used yet, especially in individual bays where waves are of maximal size. In our country, we have operating the Kisloguba Wave Station, but there are a lot of other potential sites for other stations along the coast of seas and oceans. Other interesting ideas include using wave energy in the open sea by means of floating stations. All these questions require, of course, special development.

Soviet geographers, meteorologists, and climatologists have managed to obtain enough complete data on the spatio-temporal distribution of total, direct, and scattered solar radiation incident on perpendicular, horizontal, and inclined surfaces for the entire planet and, of course, for our country. The research makes it possible to state that the practical use of solar radiation is quite feasible for many regions.

An important direction for the use of solar radiation lies in the implementation of results from studies of the spectral distribution of solar radiation into hot-house farming, taking into consideration the dynamics of the impact of different spectral parts of solar beams on vegetables and flowers. An analogous problem can be formulated in relation to treating some diseases of animals and man.

The diversity and completeness of accumulated materials on energy resources from solar radiation make it possible to start using it now, while simultaneously investigating the design of new helio-installations to increase efficiency. The participation of geographers in this field focuses on the important task of selecting the most rational sites for deploying helio-installations of different designs throughout the territory and of verifying the selection scientifically.

The use of wind resources is fairly promising for mountain regions, many flat seacoast areas, and water reservoir banks if the most efficient means are selected in every case. In a number of regions, taking into account the annual course of solar radiation and wind velocities, it is expedient to combine helio-installations with wind-energy units.

Voronov (USSR) : To preserve natural resources of the world, the creation of protected areas of different types, ranging from reservations to biospheric preserves as well as individual natural monuments, is very important. To do this, it is necessary to carry out a regionalization of the world to determine an adequate regional system. Such regional systems were created independently by M. Udvardy and R. F. Dassman of the United States. These projects were discussed at the First Congress on Biospheric Preserves in Minsk in 1983. Soviet investigators made a number of additions to these regional systems, including one for the Palearctic.

Of great importance for the preservation and rational use of natural resources is the organization of the fight against environmental pollution conducted by geographers.

Pasinich (USSR) : Intensive economic development of desert regions brings about a number of negative effects. The Sahelian zone is the most vivid example of this.

So far, accumulated aeolian relief forms have been referred to as purely exogenous phenomena. The complex analysis of geomorphological and geologo-geophysical materials has made it possible to reveal, in addition to the known connection of these forms with groundwater lenses, their link to tectonic deformations (linear-ridge forms) with synclinal and erosive lowering of the former from the surface of the impermeable layer horizon (isometric forms). In addition, it was established that such forms are associated with the local anomalies of the electrostatic field of the earth of an intensity ranging from 100 volts per meter to 1000 volts per meter.

Physico-mathematical modeling has proved unambiguously that the link between accumulated aeolian forms and electrostatic anomalies is of a regular character and can be considered as a functional connection between the structural elements of the geosystem. Of analogous character is the link between electrostatic anomalies and groundwater lenses. The formation of electrostatic anomalies is caused by absorption in the soil layer of metal and hydrogen cations. As a result of the absorption of positive charges in the soil layer, an excess of negative charges appears in ground waters. The accumulation of groundwater in structural-collectors leads to the accumulation of negative charges and, eventually, to the formation of local electrostatic anomalies.

The presence of functional links between geological structures, groundwaters, electrostatic anomalies, aeolian accumulation processes, and accumulated-aeolian relief forms allow a functional geologo-geomorphological system to be distinguished. Analysis of the structure of this system makes it possible to determine potential break points in functional links due to economic activities and, hence, disruption of the ecological balance. Groundwater and electrostatic anomalies connected with it have been shown by analysis and practical experience to be the most vulnerable. Their disappearance is, in particular, likely to be due to water loss and well drilling.

From the above description, measures to control geosystems of this type can be suggested. First, to prevent individual accumulated aeolian forms from being destroyed, it is necessary to regulate water pumping strictly. Second, pipes of nonconductive materials should be used for drilling to exclude electrical charge leakage. When developing in large regions, it is advisable to envisage the presence of protected areas extending perpendicular to the direction of dominating winds. Within the areas, natural landscapes (geologo-geomorphological systems) should be preserved to accumulate sand material drifted from the sites developed.

de Souza (USA) : We live in a world in which resources, already unevenly located geographically, are even more unevenly appropriated and consumed socially. Geographers should demonstrate that the irresponsible and irrational use of resources takes place in capitalist and socialist societies. They should also show that capitalist enterprises can contribute to responsible resource use and that the socialist mode of production can also do so. But will human societies, regardless of their mode of production, do so? To ensure that human societies do not destroy what sustains them, geographers have a responsibility to ensure the development of a world in which all people live in abundance and at peace with nature.

Cohen (USA) : An annual "early warning" system should be created by geographers for how the resources of each nation and its major regions can be most rationally used and exchanged.

Taaffe (USA) : One major need is the resolution of the differing views held by developed and developing nations on the questions of what constitutes the rational use of the world's natural resources. Other needs are: longer-range planning as regards energy; measures to cope with the consequences of the spread of an automated society to nearly all parts of the world; careful evaluation of the implications of inequalities both within and between nations, and consideration of realistic measures to reduce them.

Palm (USA) : The geographical perspective of looking at the linkages between physical and human systems is an ideal one for investigating the rational use of natural resources that follow from their investigations of the flows of materials from place to place and the likely impacts of these flows on the world ecosystem. Furthermore, geographers can carry out analyses of the world trade system that would identify sources of imbalance, and therefore targets for mitigation strategies.

Muller (USA) : We should be encouraged to "think big" in confronting humanity's environmental problems. Perhaps the international scale is the most effective one to develop multifaceted research projects --- the United States/USSR, International Geographical Union (IGU), the United Nations, and other agencies should be contacted.

Morrill (USA) : Rational use of resources, including land and water, similarly require an understanding of spatial processes and regional interdependence. Thus geography should be able to provide realistic tools for evaluation of resource use and models for resource management.

Mather (USA) : Geographers can contribute by utilizing their particular knowledge in increasing our understanding of the extent and distribution of the different natural resources. They can contribute by providing their particular understanding of how modifications in one subsystem might influence resources in another subsystem. They can, from their knowledge of resource utilization over time and in different societies, suggest ways to conserve resources or to substitute plentiful for scarce resources. They can increase their efforts to make the public aware of the fragile nature of many of our environments and the impact that one earth subsystem has on all of the other subsystems of the earth.

What measures can be suggested by geographers to prevent the consequences of an arms race?

Regarding the problem of geographers' roles in preventing the arms race and its consequences there is much communality of thought among the two scientific communities. The collective voice of the discipline is becoming increasingly loud and clear as witnessed by Soviet geographers' adoption of anti-nuclear war and arms-race resolutions at major scientific meetings in 1984 and 1985. The Association of American Geographers adopted a similar resolution in 1986. These resolutions reflect the grave concern of all geographers for the future of mankind and urge all governments to forego first use of nuclear weapons and appeal for the reduction of the risk of nuclear war in all possible ways. The responses below stress the potential of geographic research and especially international cooperative geographic research to provide scientific studies verifying the devastating results of a continued arms race and nuclear war.

Lavrov (USSR): Geographers and other scientists have already drawn the attention of society to the inevitability of an ecological catastrophe in case of mass use of nuclear, chemical, and biological weapons. It is necessary to go further and insistently remind the public that, even in the condition of stable peace, a continued arms race and the existence of the military and industrial complex prevent the solution of ecological problems and support hyperindustrialization and technocratism to the detriment of ecological ethics and morality. This is to say nothing of the waste of enormous resources. Even unused and accumulated inventoried weapons dehumanize our life and gradually destroy man and his habitat.

This task is believed to be the most important function of our science today. It can be argued that the main role of geography has always been constructive, but it would be rendered senseless in a world of "nuclear winter." Everything depends now on the resolution of a global problem --- survival.

Geography, for a number of reasons, may play a leading role in explaining to the broadest segment of the population the seriousness of the nuclear war menace. The world has shrunk. It has become very small in an epoch of jet aviation and intercontinental missiles. It has become not only small but extremely vulnerable. The catastrophe at Chernobyl has demonstrated this vulnerability only on a small scale.

These impacts can be shown clearly on a map --- the real areas of one catastrophe, likely areas of a single nuclear blow, and the much wider "interfaces" of this blow with the area of nuclear power technology. This was demonstrated by the *Atlas of Nuclear War* by the outstanding American geographer V. Bunge. Attitudes toward that author may differ, but his work was undoubtedly useful, because the map is the most visual and impressive means of reflecting such impacts, words being much less emotional. A set of such maps alone would make all speculation concerning a "limited" nuclear war senseless and remove any nonpessimistic forecasts of such a catastrophe.

Geographers (though they were not the creators of the concept of "nuclear winter" or "nuclear chaos") are able to understand and consequently clearly and vividly demonstrate what a decrease of temperature by twenty to thirty degrees means and what effects in which regions it can bring about. It is especially important because the United States and the USSR would find themselves in those regions, because they are the regions of the two greatest nuclear powers. Perhaps such maps and a detailed picture of the terrible realities of nuclear war will enable people to acquire a new way of thinking about what we speak of so much in our country. The picture of the current world --- smaller, vulnerable, extremely interdependent --- is becoming clear to scientists but is not yet clear to all people who are involved in science and are accustomed to the stereotypes of the past. Therefore, it is the duty of scientists, whether they work in Berkeley or Novosibirsk, to explain the gravity of the situation and the only possible conclusion to draw from it.

The task of promoting peace for ourselves, that is, we of the "scientific community," is no longer valid. This is verified by the statement of the Association of American Geographers (1985, 1986) and the statement of the Congress of the Geographical Society of the USSR (Kiev, 1985). I can testify that the reading of the statement of the AAG at the Kiev congress was an impressive and joyful moment. We felt that American geographers were not only our colleagues but that they shared our ideas concerning the most important question of life. And, it is not a question of ideology. One can be a Marxist or a conservative, but in both cases it is important to realize the priority of the idea of world survival and, more important, to work toward this goal.

It is, therefore, extremely important for professional geographers to leave their "ivory tower" from time to time to address "life at large," the press, radio, and television, to lecture, which we regularly do, and not to consider this to be secondary but rather essential work, and not less important than books and papers.

Another problem should be eventually solved. Our world with mass unemployment requires "sticking" to or retaining the military and industrial

complex because, it unfortunately provides jobs to millions of people. It is easy for advocates of the military-industrial complex to "prove" its necessity, which is, of course, erroneous and vicious. It is more difficult to prove that it is possible to convert modern military industries to peaceful production. It is difficult for us to recommend solutions to our Western colleagues, and besides, they know better. Too much optimism would be inappropriate. One thing is clear, that military industry in a town or region does not guarantee stable employment and prosperity. Moreover, a wider prospect for new, quite new, spheres of economic activity (particularly providing ecological stability and even improving the ecological situation in the world) must develop in the future. Still, this problem, acute as it is today, is a secondary problem. If tomorrow, figuratively speaking, the probability of an enduring nuclear-free world becomes more viable, humanity will be able to tackle the problem of removing the second threat, the ecological threat, quite seriously. And then, financial limitations that prevent dealing with this problem today will be removed and billions of rubles, marks, and dollars will be added to the "ecological sector" of economy. It will be able, in this case, to replace the unreliable and war-threatening military economy.

Is this fancy? It is only a version, an optimistic scenario of the future that may become a reality provided we work together. Then, geographers will be able to demonstrate goals for worldwide ecological planning, grand tasks --- alas, unrealistic today --- and then there will be jobs for almost everybody, and there would appear to be many fewer disasters and unemployment.

Brunn (USA): Geographers can make major contributions to the prevention of a global arms race. The contributions may be of a descriptive nature, of a theoretical model-building nature, or in the public policy arena. A variety of techniques can be used to examine related problems. These include field work to measure the effects of radiation on crop and rangeland and, water quality; social surveys to measure attitudes and perceptions; statistical methods in regional and environmental models that might be constructed; and soft and hard forecasting techniques to measure the effects and impacts of a global arms race. Most geographers should have a personal and professional interest in studying this question. Specific measures that geographers can provide to prevent the arms race and its consequences include the following: designing spatial simulations to measure the impacts of nuclear war on a nation-state or a continent; studying the environmental effects of nuclear winter on crops, climates, soils, water, and human habitations on a global and continental basis; examining the geographic effects of military economies shifting to peace economies (who are the winners and losers); measuring spatial variations in public support for a reduction on the arms race (among United Nations members, congressional votes on military-peace issues, and nuclear freeze referenda); examining the nature of the networks involved in the arms production and proliferation (industrialists, bankers, third parties) and arms reduction (peace groups); studying the global and regional consequences of arms proliferation and disarmament (specifically how economies and societies change); discussing the consequences of an arms race in our classes and professional meetings; introducing materials on the geographies of peace in our textbooks; funding research on war-peace questions; recognizing the severity of the problem and supporting individuals and groups within the profession interested in preventing an arms race and its horrible consequences; and discussing openly the above ideas with colleagues in other countries interested in the same questions and among professionals in other academic societies in our own country. While there are professional interests geographers may have in studying the impacts of increased spending and avoiding nuclear war, there are additional efforts an individual may engage in with local religious or secular groups working to the same ends.

Kondratiev (USSR): In recent years great attention has been paid to new assessments of the possible impact of a nuclear war on the atmosphere and climate. P. Crutzen (West Germany) and J. Birks (United States) were the first to discover the possible critical impact of urban and forest fires that would be produced as a result of nuclear explosions in the atmosphere.

Such fires must produce massive amounts of smoke and highly absorbing particles that will strongly attenuate solar radiation at the earth's surface by as much as 95 percent. This very substantial decrease in insolation will produce climatic cooling.

R. Turco, O. Toon, T. Ackerman, J. Pollack, and C. Sagan (a group of Americans who published their scientific results on nuclear winter in *Science*, April 1984) calculated impacts on the basis of a 1-D radiative-convective model for various scenarios of a war. They concluded that the air surface temperature drop over the continents of the Northern

Hemisphere would range up to 40° C, due to absorption of solar radiation by smoke.

Further computations made with the use of 2-D and 3-D climate models by various research groups (Lawrence Livermore National Laboratory, NASA, Ames Research Center, Computer Center of the USSR Academy of Sciences, National Center for Atmospheric Research) supported these preliminary conclusions regarding severe climatic cooling that led to the concept of nuclear winter. However, the recent NCAR calculations (Stanley L. Thompson, *Foreign Affairs*, 1986) has predicted a much weaker cooling ("nuclear fall").

The study of the global geographical distribution of climate change has shown that such change is very uneven. Some of the simulations showed that a great change of average meridional circulation can be expected. A single inter hemispheric cell of meridional circulation will form instead of the two well-known Hadley cells in the tropics. The results mean that a nuclear war would be equivalent to a global ecological catastrophe.

An important question, however, is the reliability of numerical modeling, which is difficult to assess, although it is quite clear that there are a number of assumptions which must be verified. They are, to name a few, rapid global diffusion of smoke particles (and their homogeneous distribution in the atmosphere), cloud dynamics, evolution of an atmospheric greenhouse effect, particle scavenging, and principal optical properties of smoke particles (single scattering albedo, scattering function).

Such natural analogs as the Tunguska meteor fall in 1907 and more recently volcanic eruptions (specifically, El Chichon), may be used for verification purposes, but it has been shown that their significance and degree of similarity, is very limited.

To check the validity of numerical modeling, further improvements to the models would help, and beyond that, we should analyze observed conditions that accrued after nuclear tests in the atmosphere during the late 1950s and early 1960s. High-altitude balloon measurements of total direct solar radiation at altitudes of about 30 km made by myself and G. Nikolsky (USSR) are helpful in this respect.

The balloon measurements revealed a substantial additional decrease, sometimes six or seven percent, of solar radiation in the atmosphere above 30 km after nuclear tests. Rough estimates have shown that this decrease is due to an increase of NO_2 content in the stratosphere produced by nuclear explosions.

On average for the Northern Hemisphere, if the air surface temperature decreases because of the injected NO_2 during the decade of the 1960s could reach 0.3°C, coinciding closely with observed cooling. Thus it may be concluded that climatic cooling during the sixties was due to increased NO_2 content in the atmosphere and the attenuation of solar radiation as a result of nuclear tests.

If we assume now that it is possible to scale the level of climatic cooling linearly by taking into account the cumulative power of nuclear explosions --- which could be as much as 10^4 megatons --- then nuclear war on such a scale will lead to a global climatic cooling of the order of 10° C due to the NO_2 increase.

Regarding improved modeling, an important result is indicated by 1-D calculations made by K. Kondratiev, N. Moskalenko, and S. Gusev (USSR), who carefully considered such factors as the evolution of the greenhouse effect from a prolonged burning process and the vertical distribution of greenhouse components. These calculations revealed climatic warming initially, followed by subsequent cooling and then warming again.

The scale of disturbances of atmospheric chemical composition as a result of multiple nuclear explosions is so enormous that, undoubtedly, it will lead to strong global climatic change and ecological catastrophe. It is highly probable, however, that the principal cause of the catastrophe is not just "nuclear winter" but an even more serious strong climatic instability.

The multidisciplinary nature and global scales of the phenomena involved make it necessary for geography to play a more active role in the solution of the problem of the impact of nuclear war on the environment.

Shcherban (USSR): Geographers can provide objective facts on the impact of military operations on nature, population, and economies of many countries. In Europe, World War I covered a territory over 200,000 km^2 and World War II about 3,300,000 km^2. The area occupied by Japan in 1942 was 6,000,000 km^2, with a population of 400,000,000 people.

During World War II in the Soviet Union alone, 20 million hectares of forest were felled or damaged. During the Vietnam War from 1961 to 1972 , 568,000 hectares of forest were fully destroyed, and 5,600,000 hectares were drastically damaged through the use of herbicides. About 364,000 hectares of agricultural crops were also destroyed.

During the war against the fascists, the Soviet people lost 20 million people. Every participant in the war had friends who were killed on the battlefields or who lost their health, and received first-hand knowledge about wounds and impressions of the war. Soviet geographers can provide newspaper, journal, and cinema accounts showing the state of forests, agricultural land, towns, and villages in areas occupied by fascist troops and areas in the battlefields of the war --- Leningrad, Stalingrad, Moscow, Sebastopol, Odessa, Kiev, Minsk, and many others. A mere comparison of cinematic materials and statistical data on any of the above areas for the pre-war period and immediate post-liberation period gives a vivid picture of the disastrous character of wars and warns us about the inadmissibility of the arms race.

Demko (USA): Geographers can and do make many contributions toward averting the arms race and other related issues. A number of American geographers have done excellent studies of the potential environmental and regional impacts of nuclear war --- the geography of nuclear winter. Others have studied a wide range of political-economic issues relating to world tension from a geographic perspective. Of course, a great number of significant geographical studies can be done by geographers on issues ranging from law of the sea disputes, to the geography of terrorism, to the diffusion over international space of political influence via coercion, economic pressure, and force of arms. Geographers, of all scholars, should interact, collaborate and cooperate on such studies in order to increase their scope, scale, and impact on national governments and international bodies.

There are also some things geographers should not do. They should not conduct their research to support chauvinist or politically motivated goals. Hewing to party lines and coporate interests is anti-intellectual and wrong. Ideologically motivated research should be as totally unacceptable as spewing insincere slogans.

Geographers from Eratosthenes to Strabo to Humboldt have served our art and science objectively, and we should do no less.

Morrill (USA): An obvious way is to measure and publicize the human and environmental consequences of the arms race and overt conflict. This approach, however, does not appear to have made much impact. It might sound silly, but I suspect that the best way is to encourage fairer, fuller, less ideological treatments of each other's societies is to encourage widespread travel and exchange. I suppose we must try to teach the public and our leaders that virtually all international conflict is still at base biological-psychological, that is, a fight for territory, and that there must be better ways than an arms race to resolve territorial disputes.

Willmott (USA): The reporting of simulation results, that unequivocally show the spatial extent and duration of a plausible variety of nuclear exchange scenarios, is a contribution geographers can make. Our role can be to provide the best information available to the governments involved.

Muller (USA): First, we need more research in geopolitics and political geography for the arms race - scholarship that is long on scholarship and devoid of ideological stances. Communication and mutual understanding is sorely needed --- perhaps United States and Soviet geographers can set an example.

Palm (USA): The greatest contribution that geographers can make to preventing the arms race and its consequences is scholarship on the environmental impacts of war, and particularly nuclear war. In addition, geographers can graphically portray the true costs of major national investments in arms to the detriment of investments in other items that should have greater priority given a particular resource base.

Taaffe (USA): In their concern for interdependence and keener awareness of the problems and aspirations of many countries other than the great powers, geographers could play a prominent role in pointing to problems of nuclear proliferation and the need for constant communication and negotiation among nuclear powers. Also, they, as well as all other scholars, should convey to the public the need to recognize that the world can tolerate a wide diversity of economic and political systems without the necessity of resorting to armed conflict to demonstrate the superiority of any one of them.

Cohen (USA): The arms race threatens to devour the assets and energies not merely of the major powers, but also regional powers and the developing world. The production and exchange of arms has spatial patterns that are asymmetrical with respect to productive economic exchange. Geography could focus on this distortion, both on the international and the domestic levels.

de Souza (USA): The world deployment of nuclear weaponry and the prospect of nuclear war is one of the most significant geographical issues of our time. Through their teaching and research, geographers should demonstrate to their important constituencies that the world is too small for nuclear war but is large enough for peace.

What measures are necessary to improve the international cooperation among geographers in resolving global problems?

Soviet and American geographers again form a broad consensus regarding the measures required to improve international cooperation on global issues. The range of suggestions extends from joint research projects, joint book publishing efforts, seminars, and especially the use of international organizations such as the International Council of Scientific Unions, the International Geographical Union, UNESCO, and others, for serious, international cooperative research and publication projects. Many American respondents argue for collaborative research that would bring geographic scholars together in genuine ways to solve problems of mutual concern.

Kotlyakov (USSR): During the past thirty years the international scientific community has carried out such major international programs as the International Geophysical Year, the International Biological Programme, the International Hydrological Programme, the International Programme for Studying Climate, and so forth. All of them, however, focused attention on individual components of the natural environment --- the atmosphere, the biosphere, the hydrosphere, and the lithosphere.

The creation of the *Atlas of Snow and Ice Resources of the World* by Soviet scholars serves as an example of a national project widely supported by the world scientific community. Soviet glaciologists managed to develop over eight years a new geographic-cartographic work that revealed heretofore unknown characteristics of the Alpine and polar regions of the earth necessary to understand the evolution of the landscape mantle of the whole planet.

Beginning with the early 1990s, the International Council of Scientific Unions (ICSU) will carry out an unprecedentedly large-scale international project --- IGBP. The objective of the project is to study interconnected problems of the geosphere and the biosphere of the earth to clarify the causes of changes, those natural as wel as those due to the interference of man. Understanding these causes will make it

119

possible to realize our main task --- to forecast the future of the natural environment as the basis for the further progress of humanity.

This large international program covers a wide range of problems of an animate and inanimate nature, the evolution of different spheres of the earth, man's impact and his interrelations with nature - interaction of society and environment. Of primary importance in the program is geographic syntheses of individual geospheric and biospheric investigations, the interaction of nature and society. Under the geosphere-biosphere program, the directions and trends in social and economic processes will be studied, and the differential comparative-geographical approach should be followed. The interaction of the ocean and atmosphere cannot be studied without taking into account the role of active energy zones, and so, global geographical problems cannot be understood without regionally differentiated geographical phenomena.

S. A. Evteyev (USSR): International cooperation among geographers at interstate and local levels is fairly diversified. It is carried out on both bilateral and multilateral levels. But, most important is cooperation within the framework of the International Geographical Union and its various commissions, working and research groups. Many geographers also take part in the work of other, more focused problems via international organizations such as the International Cartographic Association, the International Association for the Study of the Quarternary, the International Commission on Snow and Ice, and the International Union of Geodesy and Geophysics.

The International Geographical Union (IGU) and International Geographical Congresses (IGC) held every four years provide a communication venue for the international geographical community to exchange views and data, and to develop methods and use them in investigations in different regions of the globe. All these are positive aspects in the work of the IGU and IGC.

One of the main drawbacks in the work of the IGU is the fact that the IGU has not set itself, and hence the world geographical community, a unique goal that could unite the efforts of multiple commissions, working and research groups. This does not imply complete subordination, as every subdivision has its own clearly stated tasks. It is important, however, to ensure the participation of all subdivisions, in terms of their experience and competence, and to solve common problems set by the IGU. In addition, this would make it possible to strengthen ties between IGU commissions and groups, and render them consistent (at present many operate sporadically).

The participation of geographers in the solution of global problems --- global ecological problems primarily --- should, in our opinion, be a basis for uniting the IGU. More specifically, we are speaking of the participation of the IGU in the international program of the International Council of Scientific Unions (e.g., the ICSU Geosphere-Biosphere-Global Change Programme).

The IGBP envisages the development of forecasts of changes in the biosphere on a global scale based on studies of earth history and analyses of solar-terrestrial links and the current functions of ecosystems of different taxonomic ranks up to the global scale, as well as the biosphere as a whole, given current and predicted anthropogenic pressure.

Such work can be carried out provided at least two approaches (in which geography has traditionally been competent) are undertaken: (1) complex territorial research into ecosystems of individual regions and (2) study of the territorial distribution of productive forces in relation to natural conditions and the natural potential of a particular area, including its labor supply.

Such an approach would both ensure the optimal implementation of the new international multidisciplinary program, at least for the present decade, and make it possible to consolidate the efforts of the IGU and its subdivisions to solve current global problems, raising modern geography to a new level.

Annenkov (USSR): In addition to reconstructing programs and organizational forms of international contacts among geographers, it is necessary to reconstruct geographers' consciousness. First, we must replace the tradition of rivalry between scientific schools with cooperation to resolve scientific problems.

Ideological differences are caused by differences in research methodologies as well as by different social ideologies, views on society and paths of development, values, and the purpose and program of activities in a society formed in different social conditions. Prospects for international cooperation of geographers depend largely on the goals set by the participants, that is, solving scientific problems or "extra-scientific" interests.

Let us take an example of geographers of the United States and USSR. In these countries, major scientific schools were formed in different branches of geographical science. In the United States. geography is considered a social science and in the USSR it belongs to the earth sciences. The history of contacts between geographers of the two countries has evolved through different stages and affected changes in the social history of these countries and in world of socioeconomic relations ("cold war" periods, "detente," "crusades against communism," periods of the "Cult of the Personality"). The language barrier has been to some extent diminished by the publication of the magazine *Soviet Geography* in the United States.

In the 1970s, detente, development of personal contacts between geographers of the two countries, and a variety of other factors contributed to changing the forms and, partially, the content of ideological discussions between geographers of the USSR and the United States. The focus of debate shifted to methods and theoretical concepts. The atmosphere of personal discussions also changed. All this had a beneficial effect on cooperation.

In order to increase cooperation and communication between geographers of the United States and the USSR, serious joint works are necessary. It is not collections of papers that are needed, but rather monographs jointly prepared along mutually devolved plans on subjects important for geographers of both countries. Work on books concerning the complex study of global change of the geosphere is a step in this direction. If we manage to put scientific problems first (e.g., joint development of global problems) along with those of general human importance (strengthening mutual confidence among the nations to curb the arms race), we will succeed.

de Souza (USA): There is a need for more East-West and North-South contact among geographers. American geographers need to become far more knowledgeable about other countries if they are to help solve global problems. An expansion of links between geographers can be fostered through more bilateral exchanges and by greater participation in the activities of the International Geographical Union. Sending research and teaching materials from rich to poor countries can also help to improve international cooperation of geographers and may also help to solve some global problems.

Cohen (USA): Joint scholarly research, student exchange, and mutual access to national data useful for research purposes are obvious measures to enhance geographic cooperation. Perhaps task forces or teams of specialists could be organized to make themselves available as an integrated resource to deal with problems that are jointly defined by parallel ministries, departments, or commissions in two or more participating countries.

Taaffe (USA): More projects of the sort being suggested here would seem desirable. In the future, perhaps such projects could be devoted to a more concentrated treatment of selected substantive problems by experts with varying disciplinary viewpoints and ideological orientations. The need for more international cooperation among industrialized nations in establishing policies and priorities in foreign aid might be an example of useful cooperation, as might problems related to such things as technology transfer or the role of multinational corporations in economic development.

Palm (USA): Geographers should work closely with their respective academies of science and also in international associations such as the International Geographical Union to form panels to work on global problems. Such work is currently underway, of course, and should be widely supported by professional associations in individual nations. International seminars are useful in bringing geographers together with common research interests who can, working in concert, make recommendations to resolve global problems such as those reviewed here.

Brunn (USA): It is imperative that geographers interested in studying and resolving international problems begin to engage in cooperative efforts. We must discern whether there is indeed an interest among scholars in respective countries in studying global problems. This question can be answered by knowledgeable leaders in the countries interested in the question who already are part of some professional society or network. We must identify a small group of knowledgeable and respected geographers in each country who have professional and personal interests in understanding and resolving global problems. These geographers should represent the spectrum of current research being conducted, that is, human, physical, and human-physical. There should be an opportunity to exchange views in writing about some paramount questions relating to geography and global problems. The queries raised in this project are ideal in this regard. These responses must be shared with like-minded colleagues in other countries. There should be a series of follow-up questions by the participants that address issues not raised in this volume. Questions might be asked about priorities in research funding for addressing global problems, the publication of studies having a global perspective, the resources (human, equipment, etc.) needed to study global problems, and specific strategies to encourage a renewed emphasis on follow-up questions.

The participants should be encouraged to react to the responses of colleagues in other countries. There should be some pairing of international scholars in order to develop and promote closer working relationships. For example, the pairs or triads may be climatologists, industrial geographers, cultural geographers, or political geographers. These individuals need to be encouraged to correspond further about mutual interests, exchange publications, and investigate possible mutual research projects. Participants need to convene at some time and place to share concerns about global problems and geographers' responses to resolving them.

The agenda of such a conference could include position papers by geographers in all countries participating in the above project and opportunities for small group meetings to exchange views and to investigate cooperative (international) research projects. Among the items that might be discussed are funding for international research, field work, published and unpublished data sources, and joint-publications. The results of the above conference, that is, publishing papers and summarizing the discussions, need to be disseminated to a larger group of professional geographers.

Muller (USA): Let us get together more often, develop a meaningful and deepening dialogue, and learn to trust each other. A series of meetings should be scheduled and our work should be exchanged more intensively. Joint research, in my opinion, would be a productive way to lower some of the barriers between United States and Soviet geographers.

Morrill (USA): Ultimately, there is no substitute for scholars working together. International congresses are fine but are insufficient and expensive. There must be actual international research collaboration. In addition, in my opinion the main global problems are neither environmental degradation or even the arms race, but inequality (or inadequacy) of economic and social development --- poverty, hunger, homelessness, powerlessness, and social discrimination. No country, including the United States and the USSR, is exempt from these problems internally and from such problems imposed indirectly on other countries. The main role of the human geographer is to understand how societies organize their territory, whether .well or badly,.and their relations with other societies. In all systems, the underlying issue is inequality in the power to make (geographic) decisions. The responsible geographer can evaluate the consequences of this inequality and perhaps suggest ways to alter these conditions and ameliorate the outcomes. Let's be realistic. Our systems are not going to change markedly, but we can help make them work better and more fairly, and explore ways of cooperatively assisting the less developed world.

Taaffe (USA): To improve international cooperation, we need to create a world society where mutual trust and cooperation among all peoples exist. Geographers can take a first small step in creating such a world by having more free and open exchanges, an increased ability to travel in each others' countries, to attend meetings in each others' countries without the rigorous red tape of special invitations and clearances, and to publish in each others' journals. Certain international programs where scientists from many countries have worked together to solve particular regional or worldwide problems have helped. More such global cooperative efforts would be helpful in fostering a greater spirit of international cooperation on global environmental problems. The IGBP could be one such cooperative effort.

Myagkov (USSR): First, we must establish peace without the threat of nuclear war, and, on this basis, build confidence among scientists of different countries.

Second, we must collectively decide what the global geographical problems are and find ways of resolving them. Third, by engaging in routine organizational work (e.g., through UNESCO), having regular working meetings, and conducting research, we can improve international cooperation. Confidence based on peaceful coexistence is necessary because some geographical knowledge can be used for military purposes ("ecological war"). As long as this hazard remains, cooperation among geographers from opposing military blocs will be limited.

One of the few but encouraging examples of this kind is the thirty-year-old cooperation of Soviet and American scientists in exploring the Antarctic. It has been accomplished through the exchange of scientists and information obtained by national expeditions. I happened to work twice with the U.S. Antarctic expedition. The total number of Soviet and American scientists, who have worked together is about sixty people (three times less than the number of cosmonauts and astronauts). However, we share the opinion that such cooperation is of tremendous value. Moreover, it essentially doubles the exchange of ideas and scientific efficiency of each of the national expeditions. It also engenders mutual confidence without which cooperation is inconceivable. According to one of my predecessors working in "the other" expedition, "you come as aliens for a short time and leave as friends. Let it be so in other parts of the earth! But this, I repeat, can never be achieved without a durable peace.

Kagansky (USSR): Global problems have a regional "face." Geographers from every country are carriers of specific knowledge connected with the features of their countries, including experience solving specific problems. In order to organize cooperation, it is necessary to understand the complementary character of the experience of geographical schools of different countries. An important prerequisite for cooperation, and one of its main means, is to introduce each other not only to the results of solving specific problems, but to develop common concepts applicable to each of them. While we are now generally aware of a set of common subjects that are developed by geographers in most countries, the unique achievements of each national group remains less well known. A positive step might be to publish a series of books in our country for foreign readers specifically devoted to those lines of geographical research that are unique and specific. There are many more such directions and ideas than is customarily believed.

Demko (USA): The measures necessary for international cooperation in solving global problems are obvious. Geographers must communicate across international space and boundaries and work together with no ideological axes to grind, political agendas to adhere to, and biases to deflect their work. We must learn each others' languages, compare problems, and openly and freely develop research programs and projects. We must learn to trust each other, respect each other, and have the tolerance and patience to learn about and from each other. These are perilous times, and it would be a truly significant model for others to follow if Soviet and American geographers were innovators in genuine, international, and intellectual research efforts.

CHAPTER FIVE

In Place of a Conclusion

As editors, we believe that a conventional conclusion regarding the variety of views that Soviet and American geographers have on contemporary geography and its social relevance is premature. Such a set of conclusions would be more appropriately derived from a wider international discussion among geographers. Here we limit our synthesis to some remarks that we aim particularly at nongeographers.

The evolution of geography as a discipline has been greatly influenced by the regional diversity of natural, economic, social, and cultural phenomena which has been studied by geographers in various countries. Early schools of geography in many countries at the end of the nineteenth century were characterized by their search for the "uniqueness" of places and regions and often by environmental determinism. However, as international exchange of ideas and communication gradually increased, geography acquired its global image as a discipline searching for spatial order and studying the evolution and impact of spatial interactions, and interactions between societies and the environment.

The distinctiveness of the views expressed in the responses in this volume are conditioned by the fact that we are in a transitional stage in the evolution of geography. Today, there are many similar geographic research trends in the two geographic communities --- from academic description to interpretation of observed phenomena to the applied use of spatial or environmental knowledge in order to forecast future spatial structures and processes and even human, spatial behavior.

In addition, many subfields such as geomorphology, population geography, and biogeography generate integrated statements and elaborations of complex problems of spatial organization of landscapes and human activities at various scales from local to global.

Although an integrative or holistic emphasis prevails in the answers, there is still some evidence of geographic specialization. Clearly, geographers in both societies employ a range of methods, from the traditional qualitative techniques to computer analysis of large and complex data sets derived from remote-sensing-related methods and incorporated in geographical information systems. These new tools of geographical research are emphasized in a number of answers from both countries, not as alternatives to qualitative techniques, but rather as welcome, rigorous additions to our arsenal of analytical techniques.

The distinctions between geographical schools in the United States and USSR may be noted in the volume, but more interesting and perhaps surprising are the common aspirations of Soviet and American geographers. There is close agreement on a number of issues.

1. In this era of technological achievement and progress most societies do not fully appreciate the value and potential of geography and geo-graphical education.
2. Geography is the most integrative discipline among the earth and social sciences and emphasizes interactions between societies and the natural environment that constantly change over space and time. A recent example is found in the work of geographers in the International Geosphere-Biosphere Programme.
3. There is a pressing need to propagate geographical thinking through the secondary education system and the mass media in order to change human spatial and environmental behavior. Understanding local, regional, and global complex problems and the ability to comprehend such problems in a concrete spatial-temporal context are essentially necessary for every educated human being.
4. There is a need to include geography in curricula of higher education and in training decision makers in industry and government. Urban and rural physical planning, resource management, marketing, environmental protection, and developing population policy are a few of many issues to which the geographical approach is relevant.
5. Finally, both communities of geographers agree that geographers must participate at the international scale to seek solutions to global problems of humanity. Examples of means to increase this type of activity can be found in Part IV of this volume. In sum, geo graphers must intensify international contacts within the framework of traditional organizations such as the International Geographical Union and with new forms of cooperation (e.g., Doctors Without Borders). Peace movements should not only be political actions, but intellectual and humanistic processes, and geographers should play a role.

Some very significant comments on the future development of geography and its social role have been made by Professor Vladimir M. Kotlyakov, a member of the Executive Committee of the International Geosphere-Biosphere Programme:

There exists some opinion in our time that such sciences as geology, oceanography, and climatology have penetrated into the essence of different geospheres, while the understanding of human responsibility for the present state and the future of the biosphere will be the responsibility of public figures and masters of the arts and literature. This view is incorrect since there is a science which is responsible for solving problems of man-environment interactions. This science is geography, and its new direction is a holistic earth science.

As the twentieth century gives way to the twenty-first humanity must solve a number of scientific problems: to determine the latent laws of an altered state of the earth mantle, identified as the "noosphere" by Vladimir I. Vernadsky, to establish critical limits of anthropogenic transformation of the visible landscapes and invisible interactions; to optimize interactions between society and nature. At the same time, there are some contradictions in our state of knowledge: contradictions between the continuing differentiation of the sciences and the need for a holistic solution to geoecological problems; contradictions between the heterogeneity of scientific information and the need by decision-makers for comprehensive refined data in order to manage concrete spatial-temporal issues. In all these cases, research and advice by geographers are indispensable.

High speed computers with vast storage capacities will make it possible, within ten to fifteen years, to create "computer atlases" that will integrate new information in a real-time regime and allow experimentation with cartographic models.

To know more accurately the spatial differences in environmental changes over the earth's surface, scientists and managers need to set up a global monitoring network of geoinformation centers. The distribution of such centers should be determined by geographers. They must also assure the design and use of geoinformation systems as well as provide users with data on optimal schemes of resource management and environmental protection for every region.

Progress achieved in the earth sciences as well as in computer technology and remote-sensing techniques makes it possible to investigate problems of synthesizing accumulated information on a global scale and developing complex models of the human environment. Complex modeling of the natural environment and its change is not simple because of the multiplicity of dimensions and interactions involved. But I do not doubt that the development of the International Geosphere-Biosphere Programme with the active participation of geographers will resolve this problem.

We live in interesting but dangerous times. Our societies and our nation-states are plagued by severe strains, environmental crises, and political confrontations. The rapidity of new developments in technology borders on that of science fiction. Nations and their political leaders hope to resolve most of the problems of humanity's survival with technological solutions as we have done in the past. We cannot, however, count on such solutions in the future.

Our arts and sciences must play a more active role in the search for solutions to a wide range of contemporary problems. We firmly believe that one of the most important disciplines to re-emerge at the end of the twentieth and beginning of the twenty-first century is geography.

As noted in the preceding questions and answers by Soviet and American geographers, geography is a dynamic, rapidly developing discipline which straddles the social and earth sciences, the arts and technical fields, and societal and natural systems. It is a science of integration and synthesis, capable of dealing with disparate, interacting systems and capable of manipulating large and complex data sets with the most advanced technology.

Geographers have adapted with alacrity to the new computer technology and are employing it in many ways. They have developed Geographic Information Systems (GIS) and related, automated spatial data systems. Computer cartographic techniques and automated geoprocessing systems are growing in importance and application in all areas of geographic research. The use of remote sensing and satellite imagery is increasingly common.

We believe it is an appropriate time for geographers in various intellectual communities to enhance and promote communication among our countries. There is ample evidence in the foregoing material to verify that Soviet and American geographers, despite different intellectual traditions and emphases, have a surprising area of overlap and agreement in their approach to problems. It is also evident that there are many differences which also allow us to learn much from one another.

We believe that our two geographical communities --- among the largest in the world --- can set a very significant precedent in international cooperation by speaking to each other rather than about each other. Geographic schools in many other countries also have much to offer the world scientific community. As a discipline we have the power to solve, or at least ameliorate, many global problems and much to contribute to other sciences in resolving issues of great importance to mankind.

This volume is but one of a number of new Soviet-American initiatives in geography. This and subsequent joint projects will serve us well and, we hope, mark a first step in a trend to encourage other geographic communities to join us.

Vladimir V. Annenkov
George J. Demko

Afterword

The editors of the US-USSR exchange of views on the art and science of geography have invited me to provide some reflections on what has been said by the two groups of colleagues. I am happy to fulfill this task for two reasons. First, for a citizen in a small country located between the big powers, nothing can be more satisfying than to take part in a cross-cultural exchange about a discipline that, to my mind, has a special responsibility for stimulating thinking about fundamental things and demands their consideration by everyone, particularly those who influence the fate of our world by their actions. Second, this is a welcome opportunity to air a number of issues that have occupied my thoughts since I became an academic geographer. I shall, however, limit my comments to questions about geography as a discipline, its ontology and relationship to other fields of study and knowledge.

In general it is hard to define an academic discipline in a few short sentences. Anyone who has been steeped in a tradition of thought over a long period of time possesses a complex network of concepts that simply cannot be lifted out for inspection as a totality, certainly not verbally and probably not in other ways either. Verbal language has developed into a sensitive instrument for storytelling. But it performs poorly when we try to use it for "structure-telling." It is particularly weak when used for communicating about matters relating to space and location. Given these difficulties, it is hazardous to say if the Soviet and American statements about geography reflect real conceptual differences or whether they are simply different ways of expressing the same underlying view. Only direct face-to -face conversations would clarify this issue. Thus, when discussing differences and similarities in the following pages, I may be mistaken due to the limitations of written texts in general and of my biased understanding of their unspoken meaning in particular.

My first impression is that differences in outlook and opinions in the various answers reveal more than shades of meaning that are related to personal styles of expression. There seems to exist a pervading dissimilarity between the two national groups of geographers. Americans gravitate strongly toward the social sciences with respect to the subject matter they have in mind, whereas the Soviet geographers affiliate themselves with the earth sciences. The specific geographic contribution to knowledge is the spatial perspective for the American and the landscape as the object of study for the Soviets. Consequently, human decision makers seem to be dealt with differently --- in the United States they are to be understood, in the USSR they are to be advised.

It is interesting in its own right to ask if the difference in emphasis reflects deeper traits in the general culture of the two nations. Perhaps in a highly mobile, free-enterprise society where it is easy to buy land, space becomes a rather abstract concept with its major expression in the form of distance to overcome. Planning, on the other hand, in combination with less mobility, defines space as area to be allocated in an orderly and focused way. In this respect the USSR is more similar to established European thinking. A landscape in such a context is not a set of Euclidian points, but rather space filled with abiotic and living matter.

At a more general level there exists a perspective common to all geographic inquiries. That perspective is the focus on the multidimensional place-dependency of phenomena and events. The centrality of this view is amply demonstrated in the present text, irrespective of national origin. Thus, geography is set apart from the sciences that try to identify universal laws. Newton looked at the falling apple and the moon but had no eye for the bird that landed on the branch and shook off the fruit (if that was what happened).

I feel that geographers have been far too modest with respect to the independent identity and importance of their outlook. Achievements today may appear pedestrian compared to physics, chemistry, and microbiology, and their discoveries of the secrets of matter and energy, space and time. What is sometimes called the "Newtonian approach" has yielded an impressive understanding of the universe, intellectually connecting its smallest and largest building blocks. But such a world-view evades the scale interval where the drama of life goes on. It does not avoid it as a source of observation, of course, but rather as a goal for integrative understanding. Disregarding for the moment those insights derived from astronomical observations and space research, all knowledge of the regularities of nature have been extracted by means of experiments undertaken in our everyday world. But it does not follow that there are straightforward connections from the universal regularities back to the specific configurations of things, events, and actions around us. Place-dependency enters the picture everywhere and creates its own consequences.

The entities of our world are as they are and exist where they do, not only because of the workings of the universal laws but also, and not less, because ours is a geographic world. Nothing --- stones, plants, animals, or people --- would have become as they now are had they not developed together within the limited space of the surface of a globe. To try to understand the drama of the spatial array and relationships of all diverse phenomena and the pecking order among them is to my mind a basic science in its own right. It is also urgent that this science become intensely cultivated because of the rapidly increasing number of humans and the proliferation of technical additions to the natural world.

A strong case can be made for a "von Humboldt approach" as a counterweight to the currently dominant Newtonian approach. Of course, it could be argued that von Humboldt's ideas have been followed by the whole family of earth sciences. These disciplines, however, have largely developed as separate undertakings with limited interactions and interrelations. In contrast, it was von Humboldt's ambition to use a holistic view and a global perspective. Only recently has a major effort been made to integrate findings from the separate earth sciences (e.g., ICSU's "Global Change" program). However, the holistic perspective is still incomplete since the anthroposphere has been left out. There may well be good intellectual and practical reasons underlying such a delimitation, but it indicates that we as geographers have much to do before it is taken as self-evident that a full-fledged von Humboldt perspective must include man.

I feel that geographers in both the United States and the USSR are moving in the direction I have just indicated. Thus, for example, "place" has recently become a key word in Anglo-Saxon writing. In this context, "place" refers to much more than just location. It is nature and people, symbols and artifacts interacting under external influences emanating from other places. Place understood in this way is a more tangible reality than the somewhat outmoded notion of region. The Soviet responses employ a different comprehensive expression for place-dependence: "geosituation." This is an attractive word because it leaves open the question of whether place-dependency is limited to the immediate vicinity or involves also more scattered linkages. My understanding of the concept is as follows. In every moment everything on earth is in a historically determined geosituation. It does not matter what we choose to look at: a cloud, a stone, a tree, a house, or a person. No one is able to free himself from his geosituation. As humans we can, of course, forget that it is there. We can think of a physical scientists who plans to perform an experiment under controlled conditions in order to identify some universal principle or estimate the magnitude of some universal constant. His mind is probably fully occupied by thinking about how to arrange the experiment as such. Nevertheless, he, or his assistants, must bring together relevant materials and equipment to a locality where they can be isolated from unwanted external influences. Thus, the isolated, simple geosituation (physicists admit that they only tackle simple, soluble problems) at the core of the experiment requires a much more extensive surrounding geosituation that can be controlled. This configuration as such is not seen as a contribution to the theory of physics, however. Rather it is part of normal practice and to a large extent taken for granted. But still, as long as it lasts, it is part of the geography of the earth, even if a minor part in most cases. On the other hand, if the discovery later on leads to the creation of a new technology that spreads through the economic system, then one new geosituation after another takes shape and --- as is well known --- frequently with unanticipated and unwanted impacts.

We are presently witnessing efforts from leaders in the basic natural sciences to convince the public that more widespread knowledge of the natural sciences among people in general and politicians in particular would lead to more thoughtful applications of science-based technologies. They also suggest that scientists should develop a more responsible attitude with respect to the consequences of their work. This is, of course, laudable but neglects at least half of the problem. Laboratory findings applied in the real world lead to geosituations that cannot be derived from an understanding of the laboratory findings. In fact there exists very little understanding of how new technologies create new geosituations, wanted or unwanted. This is an enormous challenge for geographers to address.

Places and geosituations are only parts of a larger global totality. If we want to pursue a true von Humboldt approach we must consider how to move between the small-scale of everyday life and the earth system as a whole in order to investigate and understand how the various levels affect one another. The Soviet geographers introduce with a certain persistence a concept that I think could fulfill a mediating role with respect to scales. They propose the landscape mantle as the unifying concept for geographic attention. Unfortunately, I am not able to judge how close this translation renders the original meaning in the Russian language. As it stands, however, the concept certainly comes close to what I have in mind when talking about a von Humboldt approach as a restoring counterpart to the Newtonian.

In Germany, the word "landschaft" has recently come under severe attack as a key concept in geography. There is a sense that over time the word has taken on too many shades of meaning, some of them contradictory or even ideologically questionable. The English linguistic equivalent, "landscape," is no longer very common in geographic texts. Professor S. D. Brunn's reply describes it as a specific humanistic concept of interest because of "its meanings, symbols, values." My Swedish-English dictionary defines landscape as "scenery." If that is the true English meaning, it is understandable that its use is limited.

The term "landscape mantle" seems an appropriate one to avoid the narrow visual-aesthetic understanding conveyed by "landscape." It also evades obscurities attached to the single term "landschaft." The term invites one to think of all that is present in the entire surface layer of the earth: atmosphere, biota, artifacts, waters, soil, and rocks. Presence (in places or geosituations), not just visibility, becomes the criterion, not simply a presence in neat, uniform layers, but a multidimentional mosaic characterized by both inertia and instability depending on the contacts and interactions. While the systematic sciences on the whole concentrate their attention on selected objects and relations, geography should deal with the total mix of phenomena and forces and in particular with the interfaces between them.

To speak of the landscape mantle as the object of study also clarifies geography's relation to bioecology. Ecosystems, as presently understood, become subsets of the landscape mantle since the latter also contains both natural and man-made entities that do not participate in the flow of nutrients in the biosphere. They interact with these flows in various other ways, but that is another matter.

The considerations with respect to the human world are less clear. It seems to be taken as self-evident in the responses of both the Americans and the Soviets that human action should be included as part of geographic theory. I am unable, however, to judge if man and his mind are viewed as parts of the landscape mantle or as entities apart and distinct as is customary in Western thinking. Personally, I believe that we must do what social theorists seem to be unable to make clear --- that all human action is action in tangible landscapes and must be understood as such. This perspective generates a key question: what is the relationship between human meaning and geosituations?

It would take us too far afield to delve deeper into these difficult matters. A few remarks must suffice to show the fundamental role of the landscape mantle as a concept in the von Humboldt approach. First, it should be noted that everyone's mind is filled with impressions, physical and social, picked up along his unique trajectory through the landscape mantle with its mix of things and voices. Much of the knowledge and values that emanate from experiences in daily life become taken-for-granted and unreflected matters. And, it is just this circumstance that renders them powerful forces. It is no easy task to change such entrenched forces in order to turn development into sustainability. A special difficulty to be considered is that negative consequences of human action are becoming less visible than earlier. Measures taken in one place may cause damage in quite another distant place. Even locally there may be differences. Soil erosion and polluted water are obvious, but CO_2 and freons are not.

Second, I think we must learn to understand that the landscape mantle constitutes a bounded budget framework in which only a limited set of artificially created geosituations can be accommodated simultaneously (see comments of Kotlyakov and Paschenko). This means that nothing new can come into being without destroying something already in existence, either of natural or human origin. In addition, human projects nearly always have to adapt with respect to each other. So, a serious geographic study must consider conflicts between diverging interests. In this light, critical studies of both political and economic institutions and of technologies are in order.

A prerequisite for treating the landscape mantle as a limited or bounded budget framework demands initially that we develop descriptive tools that make sure that human projects and natural geosituations, whether natural or manmade, are better treated in terms of matter, energy, space, and time. I am aware that what people have in mind cannot be understood in such crude terms, but I believe that their actions toward the landscape can certainly be.

Torsten Hagerstrand
University of Lund, Sweden

About the American Contributors

Stanley P. Brunn, professor and former chair of the Department of Geography at the University of Kentucky and editor of the *Annals of the Association of American Geographers*. He is a specialist in political, urban, and social geography.

Saul B. Cohen, President Emeritus, Queens College and New York University, former head of the Graduate School of Geography at Clark University, and the former president of the Association of American Geographers. He is a specialist in problem geography.

Anthony de Souza is the geographic liaison to the Office of the President, National Geographic Society, former editor of the *Journal of Geography*, and secretary general of the 27th International Geographical Congress. He is a specialist on Third World development issues and geographical education.

George J. Demko is the current director of the Rockefeller Center for the Social Sciences and professor of geography at Dartmouth College. He is the former president of the Association of American Geographers and specializes in population and political geography as well as geography of centrally planned systems.

John R. Mather, professor and former chair, Department of Geography at the University of Delaware, secretary of the American Geographical Society, and state climatologist of Delaware. He is a specialist in applied climatology and water resources.

Richard Morrill, professor and former chair, Department of Geography at the University of Washington, Seattle. He is the former president of the Association of American Geographers and currently director of Social Science Institute. He is a specialist on spatial statistics, location and diffusion modelling, and political geography.

Peter Muller, chairman and professor, Department of Geography at the University of Miami, Florida, and former treasurer and councilor of the Association of American Geographers. He is a specialist on urban geography.

Risa Palm, professor of geography, dean of the graduate school, and vice chancellor for research at the University of Colorado. She is also the former president of the Association of American Geographers. Her specialties are urban geography and the study of natural hazards.

Allen Pred, professor and former chair of Geography at the University of California, Berkeley. He is a specialist in historical and theoretical geography.

Edward Taaffe, professor and former chair of Geography at the Ohio State University, and former president of the Association of American Geographers. His specialties are transportation geography, urban geography, and geographical thought.

Waldo Tobler, professor of geography and statistics at the University of California, Santa Barbara, academician for the US National Academy of Sciences, and member of the Royal Geographical Society of Great Britain. His specialties are spatial mathematical modeling and graphic interpretation.

Bill Turner, professor and head of the Graduate School of Geography at Clark University. His research focuses on nature-human-society relationships, global change, and human geography.

Cort Willmott, professor and chair of geography at the University of Delaware. His specialties are land-surface processes and climate and statistical analysis of large-scale climate fields.

About the Soviet Contributors

A. D. Akimenko is candidate of geographical science in the Geography Department at Moscow State University and a specialist in economic geography.

E. B. Alayev, doctor of economics, Institute of Geography, USSR Academy of Sciences, Moscow. He is a specialist on methodology and terminology in regional social-economic geography.

V.V. Annenkov, candidate of geographical science and secretary for the National Committee of Soviet Geographers, Institute of Geography, USSR Academy of Sciences, Moscow. He specializes in historical geography and the history of geographical thought.

V.A. Bokov, candidate of geographical science, and head of the Physical Geography Department at Simferopol University.

P.I. Bukharitsin is the deputy director of the Astrakhan Zonal Hydrometeorological Observatory, Astrakhan.

Yu. D. Dmitrievsky, candidate of geographical science at Leningrad Finance and Economics Institute in St. Petersburg. He is a specialist on the economic geography of Africa.

G. P. Dubinsky, doctor of geographical science and professor in the Department of Geology and Geography at Kharkov University, Kharkov.

O. A. Evteyev, candidate of geographical science and head of the Laboratory of Complex Cartography at Moscow State University, and population cartography specialist.

S. A. Evteyev, doctor of geographical science, Council for Biospheric Problems, USSR Academy of Science, Moscow, and former vice director of UNEP, Nairobi.

A. A. Grigoriev, doctor of geographical science, Institute of Limnology, USSR Academy of Sciences, St. Petersburg. He is a specialist on lakes and satellite imagery for land-use.

G. E. Grishankov, candidate of geographical science in the Geography Department at Simferopol University.

V. L. Kagansky, graduate student at the Institute of History, USSR Academy of Sciences, Moscow, and a specialist on geographical methodology.

L. N. Karpov, doctor of economics, Institute of the USA and Canada for the USSR Academy of Sciences, Moscow.

B. S. Khorev, doctor of geographical science, Population Laboratory of the Economics Department, Moscow State University, urban and population specialist.

R. G. Khuzeyev, candidate of geographical science and docent, Geography Department at Kazan University, Kazan. He focuses on the application of mathematical methods and modeling in geography.

K. Ya. Kondratiev, academician for the Institute for Limnological Studies of the USSR Academy of Sciences, St. Petersburg. He is a specialist on the physics of the atmosphere.

V. M. Kotlyakov, corresponding member, USSR Academy of Sciences, director of the Institute of Geography of the USSR Academy of Sciences, Moscow, head of the Aral Sea Committee, vice president of Geographical Society of the USSR, vice president of the International Geographical Union, and a People's Deputy of the USSR. He is a specialist on glaciology and hydrology.

P. S. Kuznetsov, professor, Geography Department at Saratov University, Saratov.

S. B. Lavrov, doctor of geographical science and professor, Geography Department at Leningrad University, St. Petersburg, vice president of the Geographical Society of the USSR, and a People's Deputy of the USSR. He specializes in social and political geography.

A. T. Levadnyuk, doctor of geographical science, Geography Department at the Academy of Sciences of the Moldavian Republic, Kishinev. He is a geomorphologist specializing in the analysis of urban construction sites.

O. P. Litovka, doctor of geographical science, Institute of Socioeconomic Problems, USSR Academy of Sciences, St. Petersburg. He works on problems of urban regionalization.

V. P. Maksakovsky, corresponding member, Academy of Pedagogical Sciences of the USSR, and head of the Economic Geography Section, Moscow State Teachers' Training Institute. He specializes in the geography of education and the global economy.

E. N. Meshechko, candidate of geographical science and is a docent at the Brest Teachers' Training Institute, Brest.

N. S. Mironenko, doctor of geographical science and a docent in the Department of Economic Geography of Socialist Countries, Moscow State University.

S. M. Myagkov, doctor of geographical science and head of the Laboratory of Snow and Ice, Moscow State University. He specializes in natural hazards.

V. E. Nekos, assistant professor in the Department of Geology and Geography at Kharkov University, Kharkov. He is a specialist on natural resources.

M. V. Omelyanchuk, candidate of geographical science at Brest Teachers' Training Institute, Brest.

L. M. Pancheshnikova, doctor of pedagogical sciences and professor, Moscow State Teachers' Training Institute, and head of the Department of Teaching Methods. She specializes in psychological problems in educational geography.

V. M. Pashchenko, candidate of geographical science and senior scientist, Geography Section, Institute of Geophysics, Academy of Sciences of the Ukrainian SSR, Kiev. He focuses the methodology and history of physical geography.

V. G. Pazinich, senior research worker, Kiev Branch of the Institute of Geology and Exploitation of Fossil Fuels, Kiev.

V. M. Peshkov, candidate of geographical science, the Georgian Seacoast Protection Laboratory, Pitsunda, Georgian SSR. He works on problems of the geomorphology of coastal areas.

V. S. Preobrazhensky, doctor of geographical science, Institute of Geography, USSR Academy of Sciences, Moscow. He is a specialist on theoretical problems of landscape analysis.

B. B. Prokhorov, doctor of geographical science, Institute for Social and Demographic Problems, Academy of Sciences, Moscow, and head of the Laboratory of Regional Health Problems. He is a specialist on medical geography and ecology.

A. M. Riman, secretary, Kharkov Division, Geographical Society of the USSR, Kharkov.

B. B. Rodoman is a candidate of geographical science and senior researcher at the National Department of Tourist Studies, Moscow. He is a specialist on theoretical issues.

M. I. Shcherban, doctor of geographical science and professor, Geography Department at Kiev University.

P. G. Shevchenkov, candidate of geographical science and docent at Bryansk Teachers' Training Institute, Bryansk.

N. B. Shkarban, candidate of pedagogical science and senior researcher, All-Union Institute of Educational Methods, Academy of Pedagogical Sciences of the USSR, Moscow.

V. S. Silenko, Department of Geography, Institute of Geology, Academy of Sciences of the Kirghiz Republic, Frunze, and systems analysis specialist on geomorphological processes.

A. A. Sokolov, doctor of geographical science at the State Hydrological Institute, St. Petersburg, and works on water resource problems.

N. M. Solodukho, candidate of philosophical science and docent at Kazan University, Kazan, specialist on development issues.

V. I. Solomatin, doctor of geographical science, head of the Laboratory of Geoecology of the North, Geography Department, Moscow State University.

A. F. Treshnikov, doctor of geographical science, academician, USSR Academy of Sciences, president of the Geographical Society of the USSR, and a polar specialist.

A. M. Trofimov, doctor of geographical science, head of the Department of Economic Geography, Kazan University, Kazan. He is a specialist on mathematical modeling in geography.

A. A. Velichko, doctor of geographical science, head of the Laboratory of Paleogeography, Institute of Geography, USSR Academy of Sciences, Moscow.

V. E. Vikulov, doctor of geographical science, Buryat Branch of the Siberian Section of the USSR Academy of Sciences, Ulan-Ude, Russian Republic.

V. V. Vorobyov, corresponding member of the USSR Academy of Sciences, director of the Institute of Geography, Siberian Section of the USSR Academy of Sciences, Irkutsk, Siberia. He is a specialist on population and economic geography.

A. G. Voronov, doctor of geographical science and professor, Geography Department, Moscow State University, Moscow. He is a specialist in biogeography.

V. S. Zhekulin, professor and doctor of geographical science, former vice president of the Geographical Society of the USSR, and head of the Physical Geography Section of the Geography Department, Leningrad University, St. Petersburg [deceased].

About the Book

Responding to the changes taking place in the post–Cold War era, the editors of this volume have brought together more than forty distinguished Soviet and U.S. geographers to redefine geography as a discipline and to examine its relationship to other sciences and to the arts. Challenging inevitable barriers of language and of differing social, cultural, and scientific backgrounds, each contributor provides personal insight and perspective, shedding unique light onto this often poorly understood discipline.

The book covers a broad sweep of issues, ranging from the methods of geography to examples of practical work done by geographers in Russia and the former republics and the United States. The contributors explore and define advances in quantitative technique, increasingly sophisticated methodology, and the essential relationship between these changes and theory building. They also examine the application of geography in Soviet and U.S. schools as well as the demands that shifting world events are placing on the discipline.

The discussions not only reveal the individual perspectives of each geographer but also provide a unique forum for the exploration of similarities and differences within the world's two largest geographic communities. The volume concludes with an afterword by Torsten Hagerstrand.

Index